Living Well in the Age of
Global Warming

Living Well in the Age of Global Warming:
10 Strategies for Boomers, Bobos, and Cultural Creatives

Paul A. Delcourt and Hazel R. Delcourt

CHELSEA GREEN PUBLISHING COMPANY
WHITE RIVER JUNCTION, VERMONT
TOTNES, ENGLAND

Boomer Breakpoints™ is a pending trademark of Paul A. Delcourt and Hazel R. Delcourt.

Designed by Jill Shaffer.

Printed in the United States.

First printing, March 2001.
04 03 02 01   1 2 3 4 5
Printed on acid-free, recycled paper.

Library of Congress Cataloging-in-Publication Data

Delcourt, Paul A.
      Living well in the age of global warming: 10 strategies for boomers, bobos, and cultural creatives / Paul A. Delcourt and Hazel R. Delcourt
            p.   cm.
      Includes bibliographical references and index.
      ISBN 1-890132-87-X
      1. Global warming—Environmental aspects—United States.
      2. Retirement—United States—Planning—Effect of environment on I. Delcourt, Hazel R. II. Title

      QC981.8.G56 D4 2001
      646.7'9—dc21   2001017351

Chelsea Green Publishing Company
Post Office Box 428
White River Junction, VT 05001
(800) 639-4099
www.chelseagreen.com

Contents

Figures and Tables   viii

Preface   xi

Part I   CASHING IT ALL IN   1

1   Quality of Life in a Changing World   5

    The Stunning Scale of Environmental Change   6

    Crises Ahead—Why Care?   6

    A Personal Checklist   8

2   Baby Boomers   10

    The Baby Boom Generation—Who Are We?   10

    Boomer Breakpoints   12

    How Do We Boomers View Ourselves?   14

    Who Will Be the Biggest Losers?   17

    Doomed because of Boomer Demographics?   19

3   Greenhouse Warming—Somebody Else's Problem?   24

    The Greenhouse Effect   25

    Commonsense Talk About Climate Models   29

    The BAU Scenario—Is It Really Business as Usual?   31

    General Consequences of Greenhouse Warming   34

    "It's Never Too Late and It's Already Too Late"   36

Part II   LIFESTYLE DESTINATIONS   37

4   The Seaside   41

    Our Oceanfront Homes at Risk   41

    Sea-Level Rise in Deep Time   42

Sea-Level Rise in a Greenhouse World 42
How Will a Greenhouse World Affect Hurricanes? 49
Deal with It, Move It, or Lose It 58
Beachfront Homes: Paradise for Years to Come? 60

5 The Lakeshore 65
How Hot Is Hot? 66
The Resorters' Dilemma 68
Forests in Peril? 79
Boom and Bust in Great Lakes Forestry 82
A Hunter's Wild Card 86

6 The Mountains 88
Wellness in the Back of Beyond 88
Paradox of the Tourists 95
The Dreaded E-Word, Extinction 102
Elusive Ecotones 103
Adapting to a Future Mountain Scenario 106

7 The Sunbelt 107
Ecological Refugees 108
Water Wars 110
Birds, the Consummate Ecological Nomads 122

Part III FINDING YOUR SOLUTION 127

8 Ecological Survival Kit 131
Packing Your Ecological Survival Kit 131
Hard Decisions 142
Risk: Changing Perceptions or Changing Realities? 144
Financial Security 145
Social Security 146
Ecological Security 148

9 Ten Best Strategies for Living Well
in the Age of Global Warming 150
Priorities 150
Risk Takers in a Greenhouse World 153
Ecological Thrill Seekers: Adventurers 154
The Ecological Contrarian 154

The Ecological Speculator   156
The Cash Flow King   157
Comfort Seekers: Ecological Nomads   159
The Snowbird   159
The Perpetual Beach Walker   160
The Time-Share Timer   161
The Sun-Seeking Road Warrior   162
The Seafaring Vagabond   163
Stability Seekers: Year-Rounders   164
The Homesteader   165
The Sustainable Hedonist   167
Winners in a Greenhouse World   168

10 Legacy for Future Generations   169
No Park Is an Island: Corridors and Buffer Zones   170
What Can the Concerned Citizen Do?   173
The Ecological Edge   181

Notes   183

Resources   214
Reference Books   214
Web Site Resources   215

Index   217

Figures and Tables

FIGURES

2–1.	The Baby Boom generation	11
2–2.	Retirement projections for Boomers	13
2–3.	Trends in the American housing market	20
2–4.	Estimated net flow of pension and retirement assets	22
3–1.	Atmospheric concentrations of carbon dioxide	26
3–2.	Projected global temperature increases	33
4–1.	Greenhouse-world projections for sea-level rise	43
4–2.	Seasonal cycles of North Atlantic hurricanes	50
4–3.	Hurricane zones along the Atlantic seaboard	55
4–4.	Protective responses to sea-level rise	59
4–5.	Hurricane storm surge and sea-level rise in a Greenhouse world	61
4–6.	Barrier island erosion, accretion, and inundation	63
5–1.	Seasonal fish habitats in southern Lake Michigan, today and projected	74
5–2.	The lake-effect snowbelt	75
5–3.	Routes taken by major midwinter storms	76
5–4.	Snowfall in Muskegon and Lansing, Michigan	78
5–5.	Projected changes in tree distribution	83

5–6. Forest canopy changes anticipated
in the Great Lakes region   85

6–1. Good and bad visibility days, Great Smoky Mountains   92

7–1. TVA projections of annual river runoff   115

7–2. TVA projections of annual hydropower   116

8–1. Priorities for Greenhouse-world lifestyles   135

10–1. Climate-warming impacts on ecological reserves   172

10–2. Subdivision plan for a "green" community   179

TABLES

4–1. The Saffir-Simpson scale of hurricane intensity   47

8–1. Greenhouse-world lifestyle chart   133

9–1. Ten strategies for Greenhouse-world winners   155

To our parents,
Eleanore and Bill Roach
Doris and Steve Delcourt

and to our children,
Michelle and Steffi Delcourt

Preface

We are poised on the brink of irreversible planetary change. The exponential growth of human population and industrial capacity are pushing the earth beyond its natural limits. Our human activities contribute to atmospheric pollution and greenhouse-like global warming, a profound change that will alter the earth's environment not only for the near future but for as long as the next ten thousand years. These pervasive environmental impacts threaten the long-term integrity of natural ecosystems: they speed up the already high extinction rates of plant and animal species, and they alter the sustainable balance among regional environments and human communities. We ourselves, in other words, are undermining the planet's ability to sustain us. Global warming is not just an abstract concept. Regional environmental changes will have important repercussions *within the communities where we live today, within most of our lifetimes.*

Our message is straightforward. As academic scientists, we believe it is true and important. But the resources we work with on a daily basis—the data sets and summaries of many investigations, along with the analytical tools that help us peer into the future—have not been accessible to the average citizen. And so the great majority of our contemporaries do not seem aware of the picture we now see emerging.

Much of our academic research has centered on solving a series of unfolding mysteries about past and present landscapes in the eastern United

States. From our studies in Quaternary paleoecology, we know that during times of rapid climate change the ecological cards are shuffled. The result is that new combinations of plants and animals arise in response to a dynamically changing world. The most satisfying part of our work is applying what we have learned about the past to understanding the present and to predicting the future consequences of global climate change.

We started writing *Living Well in the Age of Global Warming* after beginning to consider our own options for the future. We realized that planning for our lifestyle after retirement depends not only on our philosophy of life but also on the economic and ecological settings in which we'll be living. We wish to find ourselves in touch with nature, as we do each summer at our cabin on the northern shore of Lake Michigan. But we are also aware that change is an integral part of nature. We stand at a significant crossroads in time where human intervention is having an accelerated impact on environmental change. The consequences will be significant, for people of our generational cohort, our children, and our grandchildren. In talking with our summer neighbors up north, as well as our winter neighbors down south, we realized that there is widespread public interest in these issues. However, most people don't know how to learn what they need to know to plan ahead for the eventuality of environmental change. They don't have a context based upon understanding the past, and they don't necessarily understand the basis for predicting the future. We saw a need to bridge this gap.

We are not alone in our view of the Greenhouse-world future; we are simply adapting our message for a broader audience. Those of us who have spent our careers within the halls of academia write up our findings for a small community of scholars whose work is filled with jargon, mathematical equations, and projections from computer models. What we may tend to forget is that we are also part of a larger community, in our case that of the Baby Boom generation.

This book is our attempt to reach fellow Boomers with what we consider to be one of the most important messages of our generation—the prospect of worldwide changes in climate and their cascading effects on the environment, on the balance of natural biological communities, on today's human communities, and on *life-changing decisions that each of us must make*. As we enter the new millennium, global climate change is a reality. We Baby Boomers face this new Greenhouse world as we also are passing through a series of decision points that will result in lifestyle changes for many of us, especially when we consider retirement. These **Boomer Breakpoints** correspond to the passage of irrevocable threshold points in environmental change. The regional environmental changes already set in motion will have important repercussions *not only for the communities we live in today, but also for those to which we choose to retire.*

The dual specter of imminent global warming and the unprecedented influx of Boomers to retirement communities between the years 2010 and 2030 will seriously challenge many regions targeted as desirable destinations. Communities already suffering the effects of global warming will be stressed even more by the arrival of retiring Boomers as they exceed the limits of their natural and social infrastructures. In their quest for balance and sustainability (of both comfortable lifestyles and natural resources), many communities are already attempting to control the pace and density of future development. They aim to maintain the essential qualities of the good life so ardently pursued by retiring Boomers. We champion the view that both concerned individuals and communities must also plan for **ecological security**. They must broaden their sensitivity, first to recognize and then to adapt to the reality of environmental change that is now upon us.

For some individuals, that move toward greater ecological security will involve lifestyle changes. All people make choices in the quest to capture dreams. Lifestyle choices are intensely personal decisions that vary widely among us. At one end of the spectrum is an earth-friendly, "green" choice of voluntary simplicity, based on a philosophy of living with minimal environmental costs while seeking to enrich the nonmaterial aspects of life—such as spending more time with family and friends, enjoying nature, doing volunteer work. At the other extreme is an entirely materialistic mode of life based on the accumulation of wealth and heedless consumption of energy derived from fossil fuels. Most of us fit somewhere between these two alternatives.

This book is intended as a self-help guide to making appropriate decisions about future lifestyles in a Greenhouse world. It will be a world with more variable and stormy weather, which will place many of the most desirable retirement communities at risk for weather-related disasters. It will also be a world in which many populations of native plants and animals will face extinction as exotic species, alien invaders, flourish. Community infrastructure of clean water and dependable electric power will become vulnerable to changing climates, and real estate values will track the shifting position of ocean beachfront and of lake shorelines. We suggest ways that Boomers can adapt to such future environmental changes and cope with increasing risks by planning ahead intelligently. Yet the information we present is pertinent not only to Boomers but also to Generation Xers and Millennium Kids, who are following closely upon our heels and who will inherit the legacy of a Greenhouse world.

Throughout the book, we present state-of-the-science briefs on global climate warming—the scientific consensus about what is known, what still remains uncertain or unknown, and the best-reasoned scenarios available for projecting environmental change into the twenty-first century. It is our premise that we all can use this eco-futuristic perspective to plan prudently for environmental risks. In so doing, we can stack our odds for accommodating, adapting to, and even profiting from the dynamically changing Greenhouse world. We can make intelligent decisions *now* to shape the quality of our lives.

Our future rides on whether we use this critical knowledge or, by default, ignore it. In the next few decades, Boomers will sort themselves into two basic groups: **Greenhouse Winners** and **Greenhouse Losers**. By planning strategically for environmental change, Greenhouse Winners will buffer themselves from ecological vulnerability. Based on their level of risk tolerance, they will become ecological thrill seekers, or comfort seekers, or stability seekers. In sharp contrast, Greenhouse Losers will include those who fail to plan ahead. Unprepared to cope with environmental change, Greenhouse Losers will become either ecological refugees or stranded survivors. We advocate that Boomers continue to plan for their physical security as well as their fiscal security, and that we become as mindful of our ecological security as of our Social Security. In so doing, we can learn to live well in the Greenhouse world.

To that end, we offer a framework not only for understanding the troubling dilemmas that lie ahead but also for making decisions. Whatever generation you belong to, and whatever you select as your particular geographic destination and lifestyle, whether it involves working and raising a family or retirement and relaxation, we offer our view of the ten best strategies for living well in the age of Global Warming.

In this information age, the ever expanding resources available to the average citizen on the Internet give us an additional opportunity. We have sought in this book to link concerned citizens with up-to-date information sources that can help guide their decisions. We include many references to helpful Web sites, and a comprehensive set of links to useful Web sites and other resources can be found at our own Web site, which you can access at http://www.boomerbreakpoints.com.

We thank the many special people who have encouraged, cajoled, questioned, and sharpened the ideas we express in this book. Alan Berolzheimer, our editor at Chelsea Green Publishers, challenged us to expand the scope of the book, both in considering lifestyle alternatives and in spanning the generations. Barbara and Dan Bauer, Maggie and Louis Dula, Dan Delcourt, Sandy Echternacht, Duskin Fairchild, Linda Green and Steve Ferguson, Mary Ann and Jerry Garman, Judy and Tom Jones, Annie Krueger, Dale and Jim Liles, Jeannette and Jerry Mansfield, Melinda and Bill McCoy, Gary McCracken, Audrey and Larry Mellichamp, Phyllis and Dan Morse, Diane and Mike Neal, Susan Elder and Eric Roach, Janet and Charlie Scheer, Mary Tebo and Daniel Simberloff, Joann and Jack Tripp, and Stephanie and Mike Vogel either read portions of this manuscript, made helpful suggestions for topics to include, or represented role models for us while we have been contemplating future lifestyle alternatives.

We acknowledge with great appreciation the work of our many colleagues, both in academia and in government agencies, who have devoted their professional careers to understanding climate change and its implications for a future Greenhouse world. Their work is the primary basis for the projections we use in this book, and we include specific citations to their research findings in the Notes.

We especially thank our two Millennium Kids, Michelle and Steffi Delcourt, for their patience while we prepared the final draft of this manuscript, and for their irrepressible optimism as expressed in their frequent dot-com generation chorus of *"It's a Greenhouse world, after all!"*

Paul A. Delcourt
Hazel R. Delcourt

Part I

CASHING IT ALL IN

CASHING IT ALL IN

Every eight seconds, somewhere in America, a Baby Boomer turns fifty. That's a bouquet of black balloons for ten million Boomers every year. Most of us middle-aged American Boomers are preoccupied with how we will keep up with the day-to-day cost of living and the skyrocketing prices of real estate. Near-term concerns about cash flow dominate our thoughts as we just try to make ends meet. We wonder whether we can afford the luxury of even dreaming about life after retirement, let alone being able to choose the lifestyle and geographic area we prefer.

Yet at the same time many of us are hoping to cash it all in while we are young enough to enjoy many years, and even decades, of quality time. This means that we Boomers are also beginning to sock away our cash in unprecedented amounts in tax-sheltered retirement plans. We are becoming aware that retirement planning means worrying about investing—deciding how much to invest, how soon, and in what stocks.

But what are we investing for? What is the target to track? What alternatives will there be in the new millennium for us to retire to? How, for example, will we choose the right location? Can we blindly trust the glossy brochures and slick Web pages designed to lure us to seaside, lakeshore, mountain, and sunbelt resorts? Will those properties offer us the quality environment we crave for our active lifestyles? Right now is the right time to think creatively about the environmental repercussions of our location and lifestyle choices.

Many Boomers may soon wake up to the sudden realization that we are now living in a Greenhouse world. Earth's climate is increasingly being affected by global warming. Whether we are aware of it on a day-to-day basis or not, however, the environment is changing dramatically around us. Changing weather patterns are already affecting water quality and environmental stability.

Global greenhouse warming is now a reality, one that the Baby Boom generation will have to face. We can no longer assume that our ideal retirement location will be the same as we see it today by the time we are ready to live the good life. For example, the beachfront property we invest in today may not even exist by the year 2025, much less remain intact for our kids and their children to enjoy after we are gone.

In part one of this book, we look at the unique place that Baby Boomers occupy as we enter the twenty-first century. We assess who we are and what we are aiming for. Then we examine how impending environmental changes will affect us as individuals as well as our society as a whole. In part two we explore in greater detail both prospects and potential challenges that lie ahead. We consider specific opportunities linked to your lifestyle choices, and how they will be constrained by the compounding pressures that climate change and relocating Boomers will place on them. In part three we present innovative solutions available right now to enhance your quality of life, and to gain an ecological edge that might turn climate change to your advantage. We suggest ways to protect your investments and to ensure a sustainable legacy for your children and their children. We offer this advice in the hope that you will use these ecological insights to discover your own successful strategy for living well in the Greenhouse world.

1

Quality of Life in a Changing World

A high quality of life requires not only financial independence but also ecological well-being, that is, a place to live within acceptable limits of risk for environmental change. Rather than planning a strategy for retirement in a strictly fiscal landscape, we also need to think about where we want to be in a physical landscape.

We all have our own ideas about what constitutes high quality of life, reflecting widely divergent desires and attitudes. Some folks like to travel, others prefer to stay home and garden. Some people find an adrenaline rush in "pushing the envelope" on a powerful snowmobile, in a private airplane, or while running the waves on a jet ski. Others seek solitude in cross-country skiing, kayaking or canoeing on a quiet lake or stream, riding a horse on a desert trail, or ambling along an endless sandy beach. Regardless of who you are and what you choose, your lifestyle choice has a hidden cost. Taken together, the cumulative effect of all these activities becomes an environmental burden. No matter how lightly we try to tread on the land, just by living our lives we all contribute in some measure to the gathering and processing of raw materials and burning of fossil fuels that ultimately are changing the world around us.

We can, however, be proactive. We can plan ahead for the inevitable effects of global warming. By getting in tune with environmental changes expected in the next several decades in your chosen locale, you can determine your ideal lifestyle, both now and in retirement. The key is to develop a series of environmental change guidelines that will help you to refocus your thinking. Rather than being solely preoccupied with the future of Social Security, you can also opt for ecological security by developing a personal strategy for living and retiring well.

The Stunning Scale of Environmental Change

The world's cultures have developed on a planet that is ever changing. Climate has always been dynamic over a wide range of scales, from daily fluctuations in weather to millennial increases or decreases in temperature and precipitation. Many of the globally most significant changes are imperceptibly slow, well beyond the individual human life span. Past changes in global climates are of compelling interest to scientists and historians who specialize in deciphering the history of life and of human cultures. But lessons learned from the distant past can also be useful to the rest of us, informing our concerns about the *here and now* and the immediate future. Especially important as a context for change are the fluctuations in climate that occurred during the twentieth century. These changes have taken place within living memory, and they influence the way we think about our current circumstances. Yet precedents from the deeper past must also influence our choices for our future. Unfortunately, the relevant precedents for us now include some of the most significant environmental changes that have taken place at any time in four billion years of earth history.

We are in the midst of an environmental revolution. In the next 25 years, global and regional climates are going to change more than they have in at least the past century. In the next 100 years, typical global temperatures will be higher than at any time in the past 10,000 years. In the next several centuries, temperatures will exceed those of the last 100 million years. It's time to become aware of and to face these facts. Both the magnitude and the pace of global warming are sharply accelerating *beyond what humans have ever experienced*. In the new millennium, we are stepping into a brand-new world—a Greenhouse world driven by global climatic warming that will reshape the surface of the earth *for ten thousand years to come*.

Crises Ahead—Why Care?

This environmental revolution has been brought about largely by activities of humans, by our sheer numbers, and by our insatiable appetite for fossil energy and manufactured goods. Twentieth-century prosperity brought about by the Industrial Revolution, mass production of manufactured goods, and the knowledge-driven productivity of the information age has increased the overall quality of today's life for the average person in developed countries. This newfound prosperity has obvious costs, however. Increasingly intensive land use has led to overdevelopment and fragmented natural areas.

But some of the more subtle, less direct effects are more disturbing, because they are happening largely unseen. Human-caused climate change disrupts natural ecosystems. Because of habitat destruction, species of plants and animals are going extinct at unprecedented rates. Urbanization depletes freshwater reserves, ultimately limiting population growth in many regions. And global warming will have cascading effects on regional weather patterns. As a result, we face the prospect of prolonged droughts that will force crop zones to shift. Hurricanes and tornadoes are likely to become more frequent and destructive. Increasingly noticed across all of North America, "alien" species are crowding out native species of plants and animals, after being introduced here from other countries and continents. The competitive dominance of alien species, causing displacement and extinction of native species, is one kind of biological consequence of climate change with wide-ranging implications across the spectrum of ecosystems. These kinds of compounding effects can destabilize and restructure regional ecosystems and landscapes, decreasing the quality of human life. Taken together, environmental changes will have dramatic impacts on both recreational and retirement possibilities for tens of millions of American Boomers. This places before us an ecological imperative.

Instead of waiting for ecological catastrophes to strike, we can develop and use commonsense ecological strategies in order to adapt creatively to Greenhouse-world conditions. Capturing the good life means getting an ecological edge, placing ourselves where we want to be. Thus we can use this knowledge of environmental change as an opportunity. At the same time, we can act constructively to help preserve or restore natural balances that are threatened by human activities and global climate change. If we are aware of the problems before us, and of the potential for future degradation, we can act right now to help save native species and their habitats, and to use available resources more wisely in order to conserve them for the future. We can create personal solutions to immediate and longer-range problems within a positive atmosphere of enlightened self-interest.

A Personal Checklist

In order to define your ideal retirement lifestyle, you first need to take stock of your priorities. Selection criteria for living well in a Greenhouse world may include aesthetic, financial, and ecological considerations. Many elements may be intertwined in defining what constitutes the good life.

Developing your personal checklist requires making choices. A key to finding personal solutions for living well in a Greenhouse world is being able to predict the future availability of desirable surroundings, in the form of a beautiful physical landscape, a rich biological setting, or a comfortable climate zone—one that is buffered against increasing vulnerability to natural disasters driven by climate change.

Take a few minutes to develop a list of your top ten priorities for well-being in retirement. Think creatively about the locations or activities that are most important to you. Indulge in fantasies about painting "dancing bears" at your rustic log cabin at the lakeshore, savoring panoramic vistas at a "back of beyond" getaway in the mountains, watching dolphins play in the surf from the verandah of your seaside cottage, or stargazing from the vantage point of your adobe ranch house in the desert.

Is it more important to you to experience the seasons as they unfold at one place or to "snowbird," migrating with the birds in the spring and fall? Do you like the idea of an adventuresome and free life on the road in a recreational vehicle (RV), or do you need the security offered by owning a permanent residence? Do you like to plan monthly or seasonal "wild card" experiences at time-share resorts? If so, are you an autumn "leaf peeper" or a spring "wildflower pilgrim"? Is golfing in the sunbelt your life's ambition, or do you enjoy water sports or snowmobiling instead—or in addition? Do you seek solitude and long hikes in a natural wilderness setting? Or is a hunting or fishing retreat in the cards?

Consider also your social setting. Do you want to gain more self-reliance? Do you wish to be farther away from the growing population pressures of overwhelming and everexpanding municipalities? Do you long to clip that electrical-power link with the world, to live "off-the-grid" in a modest ecologically friendly home powered by renewable energy sources, perhaps growing as much of your own food as you need? Do you wish to shed lifestyle complexity for voluntary simplicity?

By jotting down those activities and priorities that are most important to you now, you will be able to identify vital sources for more information as you read through the remaining chapters. You may find yourself surprised at how your ideas change by the end of this book. For instance, you may begin to consider the green or sustainable approach to capturing the good life as a viable alternative, one that can help build an ecological legacy for future generations. Whichever course you elect, you can enhance your odds for success with our suggestions for the ten best strategies for living well in the age of Global Warming.

2

Baby Boomers

The Baby Boom Generation—Who Are We?

B orn between the years of 1946 and 1964, Boomers are the babies who came into a new world of opportunities and dreams following the close of World War II (figure 2-1).[1] We found our identity when Landon Jones first coined the term "Baby Boomer" in 1980.[2] The postwar birthrate for our population wave peaked between 1957 and 1961. Social and economic profiles of the Baby Boom generation, as tracked by the U.S. Census Bureau,[3] show that Boomers are today the core of American leadership, capitalizing on economic opportunities and working toward personal, professional, and national dreams. Numbering more than 78 million, fully one-third of all Americans alive today, Boomers are the decision makers. As futurist Ken Dychtwald puts it, by riding their "Age Wave," Boomers will use their sheer numbers and financial power to rule the twenty-first century.[4]

In recent books such as *The Great Boom Ahead* and *The Roaring 2000s Investor: Strategies for the Life You Want*, Harry S. Dent, Jr., presents a compelling case that we are now experiencing the greatest economic boom in all of history.[5] This economic boom will be shaped over the next several decades by the spending habits, innovative approaches, and organizational skills of the maturing Baby Boom generation. Cresting in about the year 2010, this spending frenzy will fuel what will potentially become the greatest and longest interval of growth and accumulation of personal wealth in the history of the United States. Ironically, as fossil energy literally fuels this frenzied growth, the resulting buildup of atmospheric pollution will accelerate the pace of human-triggered climate change.

Our fiscal landscape, however, appears to be changing even faster than the physical landscape around us. Harry S. Dent, Jr., predicts that the raging bull market in equities, sustained by new Internet synergies of information-age technologies, will benefit an unprecedented number of savvy, affluent Boomers who thrive on new challenges. As we Boomers approach maturity, we will then be challenged by an ever increasing array of retirement options and decisions. Soon we will have to shift away from our expansionist mind-set of an ever broadening future, consolidate our investments, and attempt

Figure 2-1. **The Baby Boom generation: live births in the United States, 1913–1993.** Supplemented by immigration, American Baby Boomers number 78 million in the year 2000. The post–World War II Baby Boom was followed by a demographic trough, the Baby Bust period that produced Generation X, then the Echo Boom of Boomers' kids coming of age in the new millennium, the dot-com generation. (Modified from William Sterling and Stephen Waite, *Boomernomics: The Future of Your Money in the Upcoming Generational Warfare* [New York: Library of Contemporary Thought, Ballantine Books, 1998])

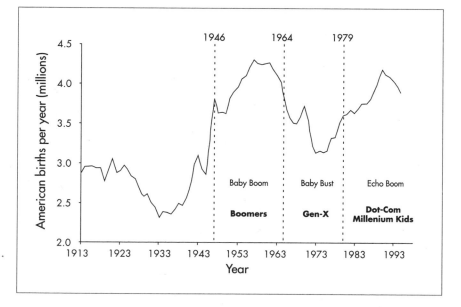

to capture our lifelong dreams. So it's time to take stock of what we want most to accomplish. As we choose, one by one, our lifestyle and retirement strategies for the twenty-first century, our decisions will drive broad demographic changes.

The key demographic changes in Boomer lifestyle, what we call **Boomer Breakpoints**, will come in the years 2010, 2025, and 2070. Three important Greenhouse-world environmental changes are projected to coincide in timing with these Boomer Breakpoints. By the first Breakpoint, in 2010, global warming will breach the typical upper limit of the temperature variability that characterizes today's weather patterns. The vanguard, the first wave of retiring Boomers, will experience the initial ecological shock wave as this first climate threshold is surpassed. The second Breakpoint, in 2025, marks the take off—the "inflection point"—for ever accelerating environmental change. The year 2025 heralds the onset of substantive and irreversible Greenhouse-world change, which will occur in the face of the peak influx of retiring Boomers. The third Breakpoint, in 2070, delimits the end of the lifetimes of virtually all Boomers, who will leave the world with at least a doubling in atmospheric Greenhouse gases over preindustrial levels. The world by then will be radically different from what it is now, as projected in the best currently available simulation models and forecasts.

In this chapter, we first explore the demographic consequences of *when* Boomers choose to retire. We then consider how their personal agendas for selecting new lifestyle goals will result in an economic meltdown of the housing and real estate market, as Boomers cash in their assets and move *en masse* to a select suite of retirement destinations.

Boomer Breakpoints

From a practical point of view, our retirement plans are constrained by our own personal philosophies as well as by our financial assets. Investment decisions made over the next two decades will in large part determine longer-term Breakpoints. Individually, our personal target dates for breaking away from the stressful workplace will typically occur at one of several predictable ages (figure 2-2). We may choose to retire early, for example at 55, when many corporate pension plans become vested, or perhaps later at 59½, when we can first tap into our individual retirement accounts (IRAs) and supplementary retirement annuities (SRAs). The first Boomer Breakpoint, at 2010, marks the decision point for the vanguard choosing to being new lifestyles coinciding with early retirement.

For many, unfortunately, harsh reality imposes yet another Breakpoint. We may have to delay the good life until we can access our mandatory government-managed, but self-financed savings. We may elect to receive partial benefits from Social Security at 62, or perhaps we will have to wait even longer, until age 66 or even 67, for full benefits finally to kick in. With such a broad spectrum of options, the demographic shift into retirement for the majority of us will happen, by choice or otherwise, at the second Boomer Breakpoint in the year 2025 (figure 2-2).

One government agency with a tangible interest in tracking the "graying of America," the Social Security Administration, has considerable computer power committed to tabulating actuarial statistical tables for Boomers.[6] According to this authoritative source, Boomer men have a typical expected longevity of 75 to 77 years; the Grim Reaper has a busy schedule ahead, harvesting most of them sometime between the years 2023 and

Figure 2-2. **Retirement projections for Boomers.** Wave after wave of Baby Boomers will retire at the ages of 55, 59½, 62, and 67, merging to form the demographic Age Wave.

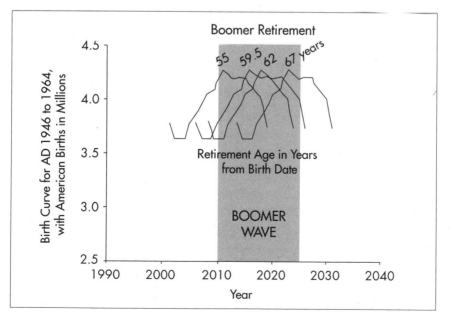

2039. Only one in 100,000 men will reach the age of 110 and live until 2056 to 2074, depending on date of birth. Women Boomers, however, will have the last word! Their typical longevity stretches to the ages of 80 to 82, with a maximum expected lifetime of 113 years. Thus most female Boomers will last until 2028 to 2044, with extreme outside limits reached by 2059 to 2077. The window of greatest opportunity for Baby Boomers to shape their future thus extends through the 2030s and even into the 2040s. Any way we look at it, asset transfer to the next generations, those Generation Xers and Millennium Kids following close on our heels (figure 2-1), will be complete by 2070. This third Boomer Breakpoint marks the longevity limit for Boomers, which will coincide with arrival of the environmental extremes projected for a Greenhouse world in which atmospheric carbon dioxide is double the level it was in preindustrial times.

How Do We Boomers View Ourselves?

We are much more than an arbitrary age group with a pecking order structured like siblings, simply by date of birth. We are certainly more than our "wealth factor"—the sum total of our current asset tally or grandiose near-term projections of wealth accumulation. We seek beyond the traditional all-American mandate of "work to consume." We are keen individualists, creative, flexible, and Internet-literate, with a green, environmentalist ideal. We strive for excellence, lasting value, personal health, and action-oriented leisure. Our personal agendas require immediate satisfaction for goals with an aesthetic flair—physical fitness, creative achievement, and the quest for peace of mind.

According to current stereotypes, several key themes paint the figurative landscape of our distinctive Boomer mind-set. First, we value high quality, and, more important, we are prepared to pay top dollar to acquire what is most important to us. With many two-career families approaching the zenith of their earning power, we have the means to buy the very best.[7] To use a pun on the jargon of "plug and play" for adding on options to our personal computers, we are prepared to "pay and play." We demand enhanced quality of life in the here and now. There's no percentage in waiting until later.

Second, we believe in the information edge. Newfound knowledge empowers our decision making, as Bill Gates says, achieved electronically at the speed of thought.[8] Simply stated, we require timely access to pertinent information to establish our priorities and then to "make them so." This is essentially an investor's mind-set: we decide our personal comfort zone for balance of risk to reward in order to bracket the envelope of real-world possibilities from which we choose. We're take-control kind of folks, experienced entrepreneurs willing to adapt good ideas and to implement sophisticated strategies for our particular game plans. Many have joined the Boomer Initiative, networking through its electronic conference center.[9]

Third, we thrive on change and require variety to spice up our lives. We continually seek new and ever broadening experiences from which to enjoy, learn, and grow. We are thrill seekers on the high wire.

Fourth, we have no time to waste. Time is our most precious commodity. We choose to *work smart*—far, far better than continuing to *work hard*. Better yet, *planning smart* up-front saves even more time. Planning smart means using all the tools available to us, and allowing a time frame in which these tools can be used effectively.

Journalist Jon Gertner views many Boomers aged 35 to 55 as members of a new class of wealth in America – the *ultra* middle class. In his *Money* magazine article, "What is wealth?," Gertner cites 1999 U.S. Census findings that today one in every eight households has an annual income of *at least* $100,000. This new threshold in family earning power, the so-called "affluence line," marks the psychological divide between households making hard choices when they spend and more-affluent families comfortably and confidently realizing their own customized dreams. To quote Gertner, "Wealth, though it can arrive with some strings attached, means access. And access, whether to the consumer marketplace or the investing marketplace, is something we have more of than ever before." Author David Brooks characterizes this segment of upper middle class as the affluent educated elite, "meritocrats" measuring their own personal success by merit-based accomplishments. In the best-selling book *Bobos in Paradise*, Brooks asserts that members of this dominant social class are *bourgeois bohemians*, or **Bobos**. Boomers, as first-generation Bobos, and their Gen-Xer offspring, as second-generation Bobos, attempt to reconcile mainsteam cultural values of the bourgeois (with worldly success measured by money, productivity, power, and position) and the counter-cultural, bohemian mindset (free-spirited attitudes for intellectual, aesthetic, and spiritual pursuits). This contemporary, hybrid culture champions a new suite of social rules for living well as cultivated individuals. This "code of financial correctness" provides upscale advice to help Bobos "convert their wealth into spiritually and intellectually uplifting experiences." For example, Bobo rule 1 states: "cultivated people restrict their lavish spending to necessities." Brooks says that rather than display one-upmanship, Bobos practice one-downmanship to subtly cultivate the ethos of voluntary simplicity. As Brooks says, Bobos "take the quintessential bourgeois activity, shopping, and turn it into quintessential bohemian activities: art, philosophy, social action." Brooks's bottom line : "a Bobo can have a big house *and* be an environmentalist."[10]

But now it's time for a reality check. Not everyone is financially affluent. Big differences exist, in both wealth and social class, among the Boomer cohort of 78 million members. Sociologist-turned-researcher of consumer markets, Paul Ray, tells us just how big, and why. In their book *The Cultural Creatives: How 50 Million People Are Changing the World*,[11] co-authors Paul Ray and Sherry Ruth Anderson highlight a dramatic cultural transition. It is not restricted to Boomers. All American adults, they say, can be grouped by their culture, values, and lifestyles. As Ray explains, these alternative ways of life simply come down to the matter of personal values. "We saw there are three competing ways of life, each with its own values, lifestyles, and views of how the world works: Traditionals, Moderns, and this new third group . . . Cultural Creatives."[12]

Traditionals, comprising 29 percent of Americans, embrace socially conservative "heartland" values centered on the family, morality, and religious beliefs. Moderns, representing a near majority of 46 percent of American adults, share a mind-set based on an honest day's work and hard-earned rewards. But they season this with a strong dose of opportunism. Moderns welcome success in the free market and relish the spectrum of goods and services created by these economic boom times. The third group, Cultural Creatives, has grown from less than 4 percent in the 1960s to over 26 percent of the adult population in 1999, numbering more than 50 million Americans today. Cultural Creatives are less materialistic and more socially concerned, advocating women's and civil rights as well as lifestyles that are ecologically sustainable. They also seek health through holistic medical practices, prefer organic foods, and pursue more meaningful lives through personal and spiritual development.

Ray's market surveys record broad cultural shifts. Since the 1960s, many children of Traditionals have adopted the Moderns mind-set. Paradoxically, many Moderns' offspring have been recruited into the rapidly expanding numbers of Cultural Creatives. Traditionals have (proportionally) lost ground. The numbers of Moderns have remained nearly constant, as those former Traditionals entering their ranks are balanced by those leaving to become Cultural Creatives. Today, three of every four American adults identify either with the relatively large (but fluid) core of materialistic Moderns or with the emerging subculture of environmentally and holistically oriented Cultural Creatives. These changing values continue to transcend other kinds of arbitrary groupings. They even cross the generational divides of maturing Boomers, Gen-Xers, and the dot-com generation of Millennium Kids.

During the next few decades, however, the personal priorities and actions of many Boomers *will converge*—from either the aggressive, entrepreneurial work ethic of Moderns, the social upscale Bobos, or the green orientation of Cultural Creatives—toward a leisure ethic in retirement (which is not to say *inactive*). This convergence will reflect our common exploration of new, creative, and challenging lifestyles. But before we can get there mentally, we have to get away from our all-consuming hang-ups about money pots and nest eggs. We have to free ourselves from those "golden handcuffs" that keep us working hard longer and enjoying life less.

Who Will Be the Biggest Losers?

Will you be a loser trapped by the system, or are you free to capture new possibilities? Social psychologists remind us of the "ostrich syndrome": some of us may think that if we stick our head in the sand long enough, we can forestall or circumvent crises. Others run full speed on the workaday treadmill just to keep up, which can be another form of mental denial. These are only temporary fixes that delay potentially unpleasant tasks, such as acknowledging and then preparing for our own and our family's future needs. In part, Alvin Toffler was right in his classic book *Future Shock*.[13] If we are unprepared and resistant to change, we may not survive our collision with tomorrow—we will be caught in an unending, futile struggle to suppress change because of our inability to adapt to the frenzied pace of the world racing past us.

Today, dominated by the Modern subculture and Bobo mindsets, American society recognizes accumulated wealth as a measure of career success. We suggest, however, that one kind of loser will be the Boomer who stalls out because of unrealistic financial goals. By not knowing when "enough is enough," one can end up perpetually tethered to a grindstone, left behind as the best opportunities for retiring well pass right on by. Alternatively, one may find oneself riveted myopically to the "green-eyed monster" of the beckoning computer screen. The economic magic of compounding interest on Treasury bonds may beguile us, enticing us either to trade them or to sit pat. Or the lure of extraordinary profits may attract some of us to the high-risk, high-adrenaline lifestyle of day trading in the stock market. The background chant is mesmerizing: "Come on, stock prices, just a little higher. . . . If I can wait just a little longer. . . . Now I have to make up for that high-tech stock loss. . . . Just not quite enough in the kitty to cash it all in yet!" Unfortunately, the last words may be the most telling.

If you find yourself tempted by any of these siren songs, it's time to reconsider options. Otherwise the epitaph on your tombstone may read, "The clock ran out before I found the time to take time for myself." The lesson here is don't deliberate yourself to paralysis or chase unrealistic goals right out of the big picture, putting off decisions about your future until the Boomer peak beats you there.

The winner will be the person who converts assets sooner to advantage, rather than later at disadvantage. Capture what you want most by acting *before* the crest of the Boomer retirement wave in the 2020s. Use this financial choreography to plan ahead for your own "optimal milieu" for quality of life. Remember that quality time (with many years remaining for good health and a stress-free state of mind) is truly the most precious commodity in our retirement portfolio.

Boomers require only one effective, bottom-line strategy to become winners. The basic rule of Boomer economics (Boomernomics) is get there *before* the rest of the Baby Boomers.[14] Place yourself *where* you want to be, *when* you want to be there, doing *what* you want. And plan for more than the merely financial. When you retire, find yourself in a **safe site**, that is, in a location sheltered from environmental hazards in an insecure Greenhouse world. Use the best available scientific knowledge to protect yourself from increasingly bigger, more frequent, and more dangerous demographic and environmental disasters. Yes, be there "first with the most." As novelist Nelson DeMille recently restated this old advice, "Second place is just the first loser."[15] The mantra of Boomernomics can, however, be carried too far. The admittedly self-serving attitude of "me first" is all too often followed by "nobody second": I've got my patch of paradise; now close the door, barricade the road, or blow up the bridge—don't let anyone else in.

Be ahead of the crowd, the Baby Boomer wave of retirement. But above all, be smart. Using an ecologically contrarian approach, critically evaluate the immediacy, magnitude, and geographic patterns of environmental changes forecast for a Greenhouse world. Eventually, though, other Boomers will notice the changes. As part of our ecological literacy, we need to realize that the hordes of people flocking to the most desirable locations will place enormous environmental and social pressures on those communities. It will be critically important to consider how those potentially endangered communities might be able to cope with the infrastructure overload in ways that are sustainable for the long term. It's not just a matter of your own enlightened self-interest; it's equally a matter of the common good. You may be among the first wave of retirees to arrive, but you will be joining an existing community with its own social history and attractive charm, perhaps also its own rooted ethnic groups different from your own. You will have to get along. Many others are certain to follow your own excellent choice, and so a high quality of life for all residents depends on both cultural and ecological sensitivity as well as on intelligent landscape planning. Boomers can use the best available scientific knowledge to act now, and to develop an ecological approach to the necessary planning. For those of us who adopt an attitude of enlightened self-interest, it *will* be possible to take control of our destiny.

Doomed because of Boomer Demographics?

The keystone in the retirement plan of many Boomers is the monetary value tied up in their primary financial asset, a residential home. Some of us may intend to stay in our present haven long after the mortgage is paid off. On the other hand, many may choose to recycle the cash value of their home in order to finance moving to another locale deemed more desirable for the golden years ahead. Should dissatisfaction with city and suburban working life become a trend, the Boomer Breakpoint in the year 2010 will launch a mass out-migration of Boomers seeking appropriate destinations for new lifestyles at the seaside, the lakeshore, the mountains, or sunbelt oases. It'll be a high-stakes guessing game for real estate marketing: *Which* new "location, location, location"?

In a recent book and magazine interview, Daniel McFadden, a 2000 Nobel Prize winner and professor of economics at the University of California at Berkeley, received national acclaim for his somber warning of the coming collapse in the real estate value of Boomers' homes.[16] McFadden predicts that there will be a demographic imperative for action when today's stratospheric selling prices plummet, after the first Boomer Breakpoint in 2010. As shown in the "Matterhorn" silhouette of figure 2-3, a chart based on McFadden's research, typical home values have already peaked in the 1980s. The projected future is a 4 percent decline in real values between the years 2000 and 2010, and a loss of another 10 percent by 2025. In all,

Figure 2-3. **Trends in the American housing market, past and projected.** Typical home prices, tracked from the year 1870 and projected through 2100, indicate that a long-term collapse of home values has already begun. (Modified from Daniel McFadden, "Demographics, the Housing Market, and the Welfare of the Elderly," in David A. Wise, ed., *Studies in the Economics of Aging* [Chicago: University of Chicago Press, 1994])

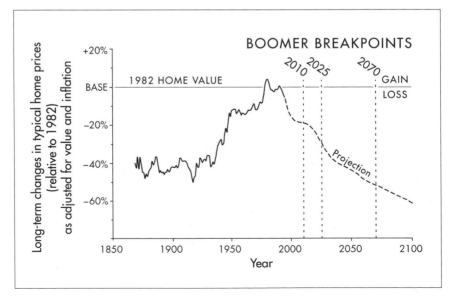

CASHING IT ALL IN

housing values are expected to bottom out after falling nearly 50 percent between the year 2000 and the third Boomer Breakpoint, in 2070. Boomers last purchasing homes in the 1980s and early 1990s face a capital loss of almost 20 percent when they cash it all in by the Breakpoint of 2010, and as much as 30 percent by 2025. Talk about corporate downsizing—how about this personal asset downsizing!

McFadden pegs falling housing values to three succeeding trends: (1) changes in demography of the Boomer population; (2) lowered housing demand; and (3) increased housing supply. The Matterhorn spike of Boomers' demand during their child-rearing years caused competition for houses in the 1980s and the crest in typical home prices, adjusted for quality and inflation. Now a long-term slide has begun. Four key factors, in turn, will shape the future demographic aspect of declining home values—accelerated Boomer retirements after 2010, their "empty nester" downsizing from large to smaller homes, their out-migration from America's continental interior to the coasts and the sunbelt, and the long-term flattening of overall population levels. This basic housing scenario of peak demand, then oversupply, is also conditioned by geography. When people sell their primary homes as they downsize, retire, or relocate, their residential location will dictate whether they realize capital losses or gains. Housing demand will tank in the rustbelt of mid-America; it should increase *only* in the most desirable retirement locations as Boomers purchase vacation homes and upscale villas within those communities. The geographic out-migration from the old, devalued place to the new, more prestigious and better-situated dream home will reflect the Boomers' perceived status of "location, location, location." Meanwhile, the ever expanding capacity of the home construction industry will continue to apply downward pressure on housing prices.

William Sterling and Stephen Waite, who are advocates of McFadden's Matterhorn scenario, emphasize the implication that we'll have some big-time losers in their provocative book *Boomernomics*. Sterling and Waite assert that the economics of the Baby Boom generation will drive a buyers' market for real estate ahead, barring extreme events such as massive immigration, war, pestilence, or famine.[17] They extend this "market meltdown" model using projections published by economic policy researchers Sylvester Schieber and John Shoven.[18] Schieber and Shoven estimate that *net cash flow* into retirement assets will peak in the year 2010 (figure 2-4). At this

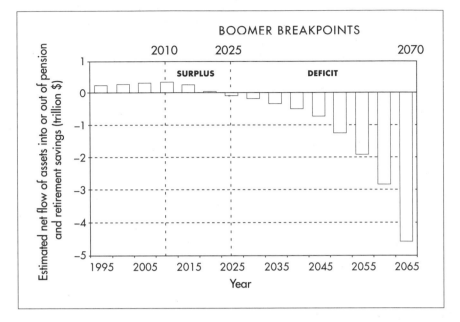

Figure 2-4. **Estimated net flow of assets into or out of pension and retirement savings.** The Boomer Breakpoint years of 2010, 2025, and 2070 mark peak inflow, reversal, and peak outflow at the end of the last Boomers' expected lifetimes. (Modified from fig. 8 of William Sterling and Stephen Waite, *Boomernomics: The Future of Your Money in the Upcoming Generational Warfare* [New York: Library of Contemporary Thought, Ballantine Books, 1998]; based on data from Sylvester J. Schieber and John B. Shoven, eds., *The Consequences of Population Aging on Private Pension Fund Saving and Asset Markets.* Center for Economic Policy Research, Publication No. 363 [Palo Alto: Stanford University, 1993], and *Public Policy Toward Pensions: A Twentieth Century Fund Book* [Cambridge: The MIT Press, 1997].)

first Boomer Breakpoint, a net annual influx of $149 billion will be socked away into American pension and retirement accounts. Following this, however, the net annual flow will change dramatically as the Boomer vanguard begins to cash it all in and literally cashes out. By the Boomer Breakpoint of 2025, the money drain will increase significantly as pension assets experience a *net outflow* of funds.[19] By the Boomer Breakpoint of 2070, each year $5 trillion more will be withdrawn by retirees than will be replenished as younger workers struggle to save for their own futures.[20]

Clearly, before we move all this money around, we Boomers are going to need an Ecological Survival Kit—with the kinds of practical knowledge and vital new insights that permit us to plot a successful path into the new millennium. The Boomer Breakpoints identify critical thresholds for near-term, dramatic changes in our generation's demography and the American economy. To formulate an effective plan for living and retiring well in the twenty-first century, we need input from a number of directions. One of them must be ecological. In a Greenhouse world with an increasingly unstable environment, we need to understand the changes in climate and environment that will correspond with the critical Boomer Breakpoints. And we need to pay special attention to changes likely to occur in the areas today considered the most attractive retirement destinations. These environmental changes, in turn, will further affect volatile housing and real estate values.

As we explain in the following chapters, environmental change is a very critical (and yet underappreciated) factor to consider in choosing a retirement community. If we ignore it, many of us may be plunging into surprising levels of real estate risk, at a time we can least afford it. As housing values plummet in some regions, ecological risks will be skyrocketing in others. Our message here is simple: Watch out, and pay attention!

3

Greenhouse Warming — Somebody Else's Problem?

n his definitive book *Global Warming: The Complete Briefing*, Sir John Houghton, scientific chairman of the Intergovernmental Panel on Climate Change (IPCC), summarized the findings of prominent scientists from 160 countries concerning the science, the impacts, and response strategies for coping with global warming.[1] As a prelude to this comprehensive synthesis, several preliminary publications had appeared beginning in 1990. And prior to the publication of those reports, Sir John was summoned in May of 1990 to Number 10 Downing Street to give an *advance* preview of IPCC findings to then British Prime Minister Margaret Thatcher. It was an important trial run for a controversial message. Sir John was told to expect a boisterous session filled with penetrating questions and rapid-fire interruptions. However, in the historic Cabinet Room, Mrs. Thatcher and her cabinet advisers heard out Sir John's presentation in awed silence. Sparked by his concluding remarks, their animated discussions reflected intense concern about global warming and its cascading series of environmental consequences. Sir John's enthusiasm was deflated, however, by the chilling comments of one senior British politician who announced that since this problem of Greenhouse warming would not become serious in his lifetime, its solution could be left for the next generation to resolve—it is somebody else's problem!

What the IPCC has shown is that, to the contrary, Greenhouse warming is everybody's problem. It is already relevant for our Boomer generation and will become even more so for our children and grandchildren. We are now beginning to witness changes in worldwide temperature that will have wide-ranging effects on physical environments, on biological diversity, and on human adaptation in the twenty-first century. Greenhouse warming will affect our choices of where we want to be in the future, what quality of life we can sustain, and what legacy we leave for future generations.

The Greenhouse Effect

What is the Greenhouse effect? In 1827, the French scientist Jean-Baptiste Fourier observed that within a glassed-in greenhouse, sunlight is converted to heat energy and trapped.[2] Fourier made the analogy between the operation of a gardener's greenhouse and the way Earth's atmosphere intercepts and traps solar energy. The atmosphere mimics the hothouse effect of an enclosed greenhouse environment.

The Greenhouse effect is responsible for maintaining Earth's temperature at an appropriate level for life to exist. In popular usage, the phrase "Greenhouse effect" applies to concerns that increases in levels of Greenhouse gases (figure 3-1), brought about by human activities, will cause changes in Earth's energy budget that will lead to significantly higher temperatures within the next century, a phenomenon also commonly referred to as **global warming**.[3]

Sunlight is transformed into heat energy not only within the atmosphere but also at the surface of the land and the oceans. Ocean currents and atmospheric winds redistribute some of this solar energy, but some is radiated back into space as infrared radiation. In the end, about half of the solar energy received by Earth's atmosphere actually reaches the planet's surface. Gases and particles in the atmosphere absorb another 25 percent of incoming solar energy, and the remainder is reflected back to space by clouds or bright areas on Earth's surface, such as the Antarctic and Greenland ice caps.[4] Why does so little of the energy escape back into space? Atmospheric gases such as carbon dioxide (CO_2) absorb heat energy, and they are responsible for trapping most of Earth's radiant heat before it can escape. Also, clouds of water vapor and other gases act as insulating blankets, intercepting some of the heat energy and reradiating it back toward Earth's surface. The balance between incoming solar energy and outgoing heat energy is thus governed by the amount of carbon dioxide and other "**Greenhouse**

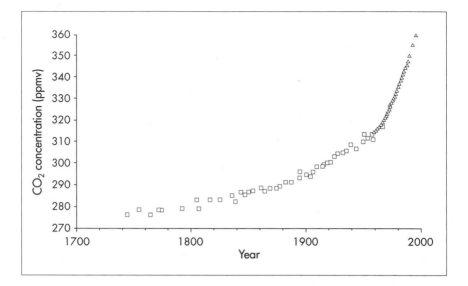

Figure 3-1. **Atmospheric concentrations of carbon dioxide.** One of the main "Greenhouse gases" responsible for global warming, carbon dioxide, has sky-rocketed in atmospheric concentration since the beginning of the Industrial Revolution. (Modified from John T. Houghton, *Global Warming: The Complete Briefing, Second Edition* [New York: Cambridge University Press, 1997].)

gases" in the atmosphere, the distribution of cloud cover, and the total area of highly reflective land surfaces such as snow and ice fields.

Greenhouse gases found naturally in the atmosphere include not only carbon dioxide but also methane, nitrogen oxide, and ozone. They are not, at natural levels, "bad" for the planet. In fact the presence of Greenhouse gases in the atmosphere is essential for maintaining temperatures suitable for life as we know it. If nitrogen and oxygen, the two predominant gases, were the *only* gases in our atmosphere, the surface temperature of Earth would be a deep freeze averaging 0°F. With current atmospheric levels of Greenhouse gases, the average air temperature is about 60°F today at the land surface.[5]

Increasing concentrations of Greenhouse gases, however, increase the amount of solar energy captured by the earth and effectively reduce the amount of radiation otherwise lost to outer space. This enhanced Greenhouse effect shifts the net balance toward a new, warmer equilibrium level. It changes atmospheric and oceanic temperatures and also alters circulation patterns, thus changing regional climates as well as the global average.

Our current dilemma has arisen because of significant increases of Greenhouse gases that have occurred since the start of the Industrial Revolution in the 1750s. Enormous amounts of carbon have been released into the atmosphere in the form of carbon dioxide gas (figure 3-1) because we have been burning fossil fuels for manufacturing goods, for transporting goods and ourselves, and especially for generating electricity. Carbon dioxide levels rise as well through deforestation of large areas for agriculture. As a result, the level of atmospheric carbon dioxide has risen from an estimated 280 ppmv (a measure of concentration expressed as parts per million by volume in the atmosphere) in preindustrial times, some 250 years ago, to current levels that exceed 360 ppmv.[6]

Another Greenhouse gas, methane, is even more effective than carbon dioxide at trapping heat energy. Methane concentrations in the atmosphere are currently rising about 1 percent per year, like carbon dioxide. But methane reacts photochemically with sunlight to form ozone, carbon dioxide, and water vapor in the atmosphere, which are themselves Greenhouse gases—making it an even more potent contributor to global warming. Airborne methane arises from the decomposition of organic matter within landfills, wetlands, and agricultural land. Methane gas is also produced from coal mining and from the industrial extraction and refining of oil and natural gas.

Another category of Greenhouse gases includes artificially synthesized compounds grouped under the trademark Freon. Many are chlorofluorocarbons (CFCs), gases first produced in large quantities in the 1960s for air conditioning and refrigeration units. As a regulated class of Greenhouse gases, these CFC compounds are now being phased out because of international concern about their effect on the ozone layer over the North and South poles, producing an "ozone hole" in the atmosphere every year. It is worrisome because ozone in the upper atmosphere acts as a shield, protecting life on Earth from harmful incoming ultraviolet radiation.[7]

Three experts on the subject of climate change, Sir John Houghton, David Gates, and Stephen Schneider, have summarized in their books the feedback effects between increases in Greenhouse gases in the atmosphere and global climate change.[8] These relationships are used to devise computerized **general circulation models (GCMs)**, which simulate circulation patterns of air and water within Earth's climate system. Once calibrated for the present-day climate, GCMs can be used to predict, within certain limits, the course of future climate changes under a variety of scenarios. For the twenty-first century, the GCMs yield estimates of average global temperature ranging from 3 to 5°F warmer than today, given a doubling of atmospheric carbon dioxide from preindustrial levels. Global warming of this magnitude should accentuate regional shifts in weather patterns, trigger widespread extinctions of tree species, destabilize natural biological communities, and disrupt the predictable supply of agricultural products, vastly complicating global politics in an already strife-filled world.

How does this jump of 3 to 5°F by the year 2100 relate to the historic pattern of temperature fluctuations? One specialist in reconstructing ancient climates from fossil evidence, Ray Bradley from the University of Massachusetts, has produced a graph of Earth's average temperature that extends back a thousand years. He found only minor fluctuations in temperature of some 1 or 2°F, bobbing around an average similar to today's. Even these modest climate changes have had devastating impacts, particularly for people living at high latitudes and on land marginally suitable for agriculture. Bradley suggests that "all our experience from the Middle Ages on is trivial, compared with what's in store for us in this century."[9] Climatologists describe Bradley's graph as the "hockey stick": typical temperatures in the near future are expected to trend dramatically hotter, flaring upward like the abruptly upturned blade on a hockey stick.

Climatologists Tom Crowley and Kwang-Y. Kim use GCM models to project Earth's average temperature over the next millennium and far beyond. Presuming that humans will use a substantial amount of existing fossil fuel reserves, Crowley and Kim forecast a spike in overall temperature between the years 2200 and 2400 that will probably reach 13°F warmer than today and possibly even climb to 23°F hotter. For context, over the past twenty thousand years, the average temperature of the globe has warmed only 9°F, between the extreme cold of the Pleistocene ice age and the modern interglacial climate. Only a few centuries into the future, the warming we are causing will be *superimposed* on the natural trend of warming between ice ages. This magnitude of a spike in average global temperatures *has not been experienced on Earth for at least the past 100 million years.* Crowley and Kim's GCM results further project that human-induced Greenhouse warming *will persist for the next 10,000 years* before natural processes reassert control over Earth's climate.[10]

Can we believe such results? Greenhouse world projections made by GCMs are the best available. Can we afford the luxury of *not* taking the futuristic projections of these climate models seriously?

Commonsense Talk about Climate Models

Computer models can, within limits, help us to predict next week's weather. Computer models that are more complex enable us to forecast future climate patterns. All of them use mathematical equations based on physical laws to simulate circulation patterns of the atmosphere and ocean—the winds and water currents that drive Earth's climate system.[11]

Climate models vary in the number of dimensions they simulate as well as in the amount of detail they include. The most simple models may calculate the average global temperature as a balance between sunlight bounced back from Earth's surface and the heat-trapping, greenhouse properties of the atmosphere—all handled as global averages. More sophisticated, three-dimensional models simulate variations in temperature with latitude, longitude, and altitude. Most complex of all are the GCMs, which are capable of predicting in three spatial dimensions not only changes in temperature but also in humidity, wind speed, soil moisture, sea ice, and other environmental variables.[12] Typically, the GCM computer simulations are run many times, using the same set of starting conditions, to determine the full range of possibilities. Today's GCM models are of high quality and very credible, and are able to simulate our modern climate closely.[13]

Differences in computer results are often tied to decisions made about the importance of feedback mechanisms. Critical feedback mechanisms include the complex role of clouds, which both cause and are caused by changes in temperature (among other things). Clouds may reflect incoming sunlight and thus be a potential cooling mechanism, or alternatively they may trap energy near the Earth's surface, hence providing a warming mechanism. Scientists differ somewhat in their opinions as to the relative significance, or weighting, of such integral components as clouds and their feedback interactions. This diversity of ideas has led to a virtual cottage industry in developing new GCMs to explore the possible roles of various linkages among air, land, sea, and life. To learn more about the inner workings of GCMs, read the report (available online) from the prestigious PEW Center for Global Change. From this document, you can discover for yourself, as climate modeler Tom Wigley says, "what is known and what is not known about the science of climate change."[14]

From the many climate models available, two GCMs are widely considered to bracket the plausible range anticipated for Greenhouse warming. If you liken Greenhouse warming to a cosmic stove top with solar energy heating up the "pot" of Earth's climate, you could think of the more conservative **GISS climate model** as the *slow simmer version*, and the more extreme **GFDL climate model** as the *full boil version*. The GISS climate model was developed by Jim Hansen and his colleagues in New York at the Goddard Institute for Space Studies (GISS) of the National Aeronautics and Space Administration (NASA).[15] The GFDL climate model was developed by Richard Wetherald and Syokuro Manabe in Princeton, New Jersey, at the Geophysical Fluid Dynamics Laboratory (GFDL) of the National Oceanic and Atmospheric Administration (NOAA).[16]

Both the GISS and the GFDL models start with atmospheric concentrations of carbon dioxide at modern levels and then simulate an increase of 1 percent per year. This produces a calculated doubling of atmospheric carbon dioxide after seventy simulated years, by the Boomer Breakpoint year 2070. Today's simulations—by GISS, GFDL, and other models—reproduce with reasonable accuracy both the short-term weather and the longer-term climate patterns and processes we have observed over recent decades. Climatologists now understand the big picture of the climate system and have largely captured how it works with their GCMs.[17]

What, then, is lacking? Why the apparent skepticism in some quarters? According to global change authority Sir John Houghton:

> Because [climate] model simulations into the future depend on assumptions regarding the future . . . emissions of greenhouse gases, which in turn depend on assumptions about many factors involving human behavior, it has been thought inappropriate and possibly misleading to call the simulations of future climate so far into the future "predictions." They are therefore generally called "projections" to emphasize that what is being done is to explore the likely future climates which will arise from a range of assumptions regarding human activities.[18]

In other words, the primary source for potential error in our Green-house-world scenarios lies *not* in our scientific knowledge or use of computer models. Rather, most of the uncertainty lies in our "guesstimates" about human actions, including, for instance, the political decisions of our governments. Will there be treaties, campaigns, or economic incentives to reduce the emission of Greenhouse gases? It's not about nature—it's about human nature!

For this, there are other models. In *Beyond the Limits,* Donella Meadows and her co-authors describe their research they conducted over the past several decades in which they used computer models to forecast global economic trends through the year 2100. Without exception, their computer simulations converge on a consensus of ever expanding population growth, increasing industrial demand for raw materials and manufactured goods, and inability to offset or curb environmental collapse. They describe a future Greenhouse world with depleted natural resources and an exponentially accelerating human population, overshooting the earth's capacity. Ultimately, the sad results are constrained by the maximum amount of land area suitable for agriculture and by the negative feedback of industrial pollution, which degrades the land's sustainable capacity. Meadows and colleagues offer several possible solutions to this dilemma. They champion an optimistic set of policies centered on making deliberate lifestyle choices that will result in zero population growth, wisely managing resource consumption, shifting toward cleaner energy sources, and enhancing technological efficiency. To ensure a high quality of life in a sustainable future, Meadows advocates replacing unconstrained growth by managed eco-development.[19]

The BAU Scenario—Is It Really Business as Usual?

What other scenarios might humans follow? Well, we already are following a classic one. Sir Isaac Newton's first law of motion states that a body at rest tends to remain at rest, and a body in motion tends to stay in motion in a straight line *unless* acted upon by an outside force. To paraphrase this natural law of physics, politicians are not likely to enact environmental programs intended to stabilize or reduce emissions of Greenhouse gases *unless* they are pushed into motion by their constituencies. Rather, they are more likely to continue business as usual (**the BAU scenario**).

The 1992 IPCC report projected different possible trends in the human-caused increase of Greenhouse gases (such as carbon dioxide, methane, nitrous oxide, and CFCs) through the year 2100.[20] Six **International Scenarios (IS)** bracketed the plausible suite of Greenhouse-world possibilities. These scenarios, labeled IS 92a (the BAU scenario) through IS 92f, incorporated global projections of population and economic growth, the future costs and availability of different energy sources (including renewable, nuclear, and fossil fuel), advances in energy efficiency, tropical deforestation, and cumulative social responses to changing political and environmental conditions. Thus, the IS scenarios attempted to project climate change given six different rates of carbon dioxide buildup in the atmosphere. In contrast, the GISS and GFDL climate models use an assumption that atmospheric carbon dioxide will increase 1 percent per year, reaching by 2070 a concentration double what it was before the Industrial Revolution.

Five of the IS scenarios presume that the absence of new control measures will result in substantive increases in future emission levels for Greenhouse gases (figure 3-2). All scenarios incorporate government policies shaped by international agreements adopted as of December 1991 to mitigate climate change. For example, global emission of the CFC refrigerant gases is expected to stop completely between the years 2075 and 2100. (International manufacture of these CFCs is scheduled to be phased out during the first decade of the twenty-first century, but existing stockpiles will continue to persist in obsolete cooling systems and hence continue for many decades to escape into the atmosphere.)

Of the six projections, the **BAU Scenario** (IS 92a) is considered the most realistic and probable vision of our future, leading to a *tripling* in annual levels of carbon dioxide emissions by the year 2100 (and thus a doubling within ninety-five years of carbon dioxide concentration within the atmosphere). This middle-of-the-road projection adopts the World Bank's estimate of a global population steadily rising to 11.3 billion by 2100, nearly doubling from 6.0 billion in 1999. As much as 94 percent of this future population growth will be concentrated within developing countries. Moderate economic growth is assumed, at an average annual rate of 2.3 percent for global gross national production (GNP), an amount that is still less than historic rates in developed countries. In this scenario, the combination of rising

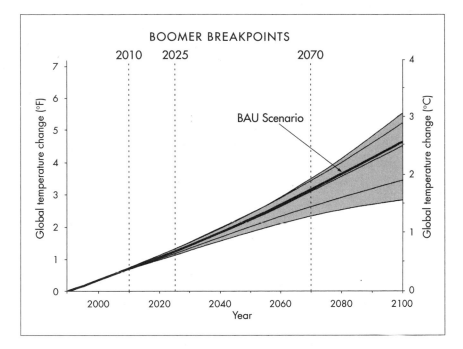

Figure 3-2. **Projected average global temperature increases for the Greenhouse world.** These projections, made by the Intergovernmental Panel on Climate Change (IPCC), display the envelope of possibilities anticipated for global Greenhouse warming. The most probable case, the business-as-usual (BAU) scenario, offers a middle-of-the-road forecast in which the curve bends increasingly upward, starting around the first Boomer Breakpoint year of 2010. That means the *rate* of warming *accelerates* all through the years that Boomers will be retiring. (Modified from Tom M. L. Wigley and S. C. B. Raper, "Global Mean Temperature and Sea Level Projections under the 1992 IPCC Emissions Scenarios," pages 401–404, in R. A. Warrick, E. M. Barrow, and T. M. L. Wigley, eds., *Climate and Sea Level Change: Observations, Projections, and Implications* [Cambridge: Cambridge University Press, 1993].)

population numbers, increasing income levels, and improving living standards will drive global energy use and Greenhouse-gas emissions.

The **Green scenario** (IS 92c) provides the most optimistic projections, including a relatively low global population growth that first rises to 7.6 billion by Boomer Breakpoint 2025, then declines to 6.4 billion by 2100. Globally, zero population growth is achieved by late in the twenty-first century. Annual emissions of carbon dioxide rise modestly by 2025, then fall markedly by 2100 to nearly one-third *below* the 1990 baseline. The minimal population growth limits the need for converting forested lands to agriculture, slowing rates of deforestation. An assumed low availability of oil and natural gas resources makes fossil fuels relatively expensive and shifts energy use toward alternative, renewable, and nuclear sources. The combination of low global economic growth (at an average annual GNP rate of 1.2 percent) and rigorous environmental regulation, with strict pollution controls, drastically reduces overall emission rates of Greenhouse gases.

The **World Out of Control scenario** (IS 92e), on the other hand, factors in rapidly accelerating population growth, projecting 9.4 billion people by Boomer Breakpoint 2025 and 17.6 billion by the year 2100. In this "bad news" scenario, the annual production of carbon dioxide expands by the year 2100 to a level *five times greater* than the 1990 baseline. Expanding populations, their need for increased food production, and the degradation of arable lands drive a continuing deforestation. This "upper-bound" scenario incorporates high economic growth (GNP 3.0 percent) with a high availability of fossil fuels, supporting higher standards of living within developing countries. Both population pressure and increased human aspirations then accelerate the pace and magnitude of environmental changes within this "out of control" Greenhouse world.

The six IS scenarios (92a through f) *all* embody overall trends of demography, economics, and pollution that produce long-term buildups of Greenhouse gases in the atmosphere. In these scenarios, the proposed annual additions of carbon dioxide into Earth's atmosphere result in *at least* a doubling, to 700 ppm, by the year 2100. These scenarios provide the spectrum of best-estimate curves for global climate warming (figure 3-2) and encompass the realistic range of possibilities for both the magnitude and the rate of global warming.[21] The message is clear—our future *cannot be* simply business as usual! Within the lifetimes of Baby Boomers, the Greenhouse world will be changing in ways and at rates for which there have been *no precedents*.

General Consequences of Greenhouse Warming

David Gates has outlined some of the general consequences of continued increases in Greenhouse gases.[22] First and foremost, we can expect an overall increase in air temperature at the earth's surface. That will cause rapid changes in landscape and agriculture. Over the next 70 to 100 years, global warming by 5°F above today's average will significantly shift the regions in which both native and cultivated plants can grow. Accompanying this widespread warming, the range between typical daily maximum and minimum temperatures will shrink, resulting in hot days and much warmer nights during the growing season. Polar regions will experience milder and shorter winters. As a consequence, at high latitudes, sea ice will thin, and areas of open marine water will enlarge seasonally. The melting of sea ice will in turn accelerate further Greenhouse warming in polar regions by as much as three times more than the global average.

Over continental regions, seasonal cooling in autumn will be reduced, snow cover will diminish, and springtime warming will begin earlier each year. Averaged over the planet, annual precipitation will increase as water evaporates more quickly with gradually rising temperatures. However, some areas, such as the drought-prone continental interior, will experience a greater seasonal contrast between wetter winters and very dry summers. The earlier melting of snow and onset of drying in the spring will lead to long-term deficits in summer soil moisture in midcontinental regions. But farmers will not be alone in struggling with the consequences. Severe summer droughts will favor more frequent and widespread fires across many landscapes.

Continental edges, or seacoasts, will face different problems. The average sea level will rise throughout the world's warming oceans because of the thermal expansion of seawater and progressive melting of continental glaciers and polar ice. Weather circulation patterns will become increasingly stormy. Trapped underneath the insulating atmospheric blanket of Greenhouse gases, the energy from incoming sunlight will provide more energy to drive the earth's climate system. This natural solar power will generate and drive powerful storms carried farther than ever by air and water currents, transferring energy from the torrid equator toward the frigid poles. Tornadoes, hurricanes, and typhoons will all become more frequent and more intense. As these powerful Greenhouse-world storms release large amounts of rainfall, flooding will become a crisis experienced repeatedly within river corridors and in coastal areas already prone to deluges.

David Gates predicts that these Greenhouse-world changes will occur very rapidly during the twenty-first century as natural thresholds are exceeded and the climate system destabilizes. Based on their computer models, climatologists agree that global warming will occur at a rate that is unprecedented in all of Earth's history.

"It's Never Too Late and It's Already Too Late"

David Gates uses this enigmatic phrase in connection with global strategies intended to minimize the effects of Greenhouse gas emissions. The sooner we take appropriate action (in this case at the international level of governments and major industries), the less drastic our actions have to be, and the more effective those actions will be. As a society, if we take actions to minimize any further acceleration of Greenhouse warming, as individuals we won't be reduced to merely developing personal strategies to cope with and adapt to extreme environmental changes. But we will still have to cope with unprecedented changes, even with the mildest possible global warming.

How, then, can individuals adapt? We can develop broader strategies and lifestyle choices that are proactive, and that contribute incrementally to reducing the magnitude of projected environmental changes as well as to softening their impact. We are stakeholders in the future. Our votes select the politicians that shape government policy. We are also stockholders who can direct our support to appropriate multinational corporations.[23] The personal choices available to us include using environmentally green building practices, shifting to renewable energy sources, recycling materials, carpooling, purchasing hybrid or electric cars, and other conservation-oriented activities. Our collective purchasing power already promotes a broader choice of products, including organically grown food offered at both grocery stores and farmers' markets.

By understanding more about global warming, and how it will affect local and regional environments, we will be able to make better choices as consumers about future lifestyles, including Boomers' choices about retirement destinations. This knowledge will be crucial in helping us adapt to the radical changes predicted for the Greenhouse world in the next few decades.

Part II

LIFESTYLE DESTINATIONS

"The weather looks a little iffy."

LIFESTYLE DESTINATIONS

Over the next century, the Baby Boom wave will be driving global economic growth, with key times of change (Boomer Breakpoints) anticipated at 2010, 2025, and 2070. The twenty-first century is also going to be a time of environmental transition for which the ecological message is now quite clear. Because of environmental changes associated with Greenhouse warming, economic life as we know it today (business as usual) will not continue. No matter what you expect or hope about future geopolitical actions that might be taken to offset or reduce pollution, Greenhouse gases will continue to accumulate in the atmosphere and will lead to unprecedented global warming. Environmental change is inevitable, and what has already been set into motion will continue for many millennia to come. It's a permanent departure from the past.

The choices we Boomers make for our retirement lifestyles and destinations must be made within the context of a rapidly changing Greenhouse world. But rather than just worrying about these changes or, worse, ignoring them, you can make viable decisions by using environmental change to your advantage. Empowered by the newfound knowledge of Greenhouse-world changes as outlined in the next several chapters, you can prepare to transform potential problems into opportunities.

In part two, we take as examples several of the most popular destination areas in North America. Chapters 4 through 7 can be used as a how-to manual to combine lifestyle decisions with a Greenhouse-savvy selection of desirable places to live during your retirement years. What do we know about how environmental changes will affect living in a particular setting? How will mass in-migration of Boomers to desirable destinations impact already established communities as they attempt to manage increasingly overloaded infrastructure capacities and traffic congestion? What strategies are appropriate for coping with those changes? We offer here a perspective on how to use ecological literacy to plan and implement a high-quality lifestyle according to your personal goals and your personal level of risk tolerance, *whatever* your geographic destination.

CHAPTER

4

The Seaside

A good friend of ours is planning to retire to an island paradise in the Caribbean. As a teacher, he is not able to afford the skyrocketing price of a beachfront villa or an exclusive home on the island summit, with its commanding panoramic views of sand and sea. His solution is to buy up the relatively cheap land midway up the hillslope now. His investment plan is to wait for Greenhouse warming to melt polar ice, then for sea level to rise and bring the beachfront to him! Is this Greenhouse strategy incredible and wild-eyed, or is it going to pay off big-time for him as an incredibly shrewd real estate speculator?

Our Oceanfront Homes at Risk

The seaside, that delicate coastal zone balanced between land and water, will respond dynamically and dramatically to climate change within our lifetimes. Already, more than half the population of the United States lives in counties situated along the twelve thousand miles of coastline. By the year 2010, rapidly expanding coastal populations should grow by 60 percent beyond their 1960 levels. By this first Boomer Breakpoint, demographers project that 70 percent of all Americans will live within one hundred miles of the coast.[1] Boomers need real answers to some very practical, very relevant questions. For instance, how much and how fast will sea level change in a Greenhouse world? How will warmer sea-surface temperatures influence the location, frequency, and intensity of tropical storms and hurricanes? Will any stretches of the Atlantic and Gulf coastlines be protected from the onslaught of major hurricane winds and their devastating storm surges? Will the federal government program for flood insurance bail us out if and when our beachfront homes are flooded by storm surge? What do we need to know about land elevations, historical patterns of coastal erosion, and building codes to construct a dream house that will last as a long-term investment, with increasing property values? But how long is long-term? Will our beach home survive the advancing wave attacks long enough for our children and our grandchildren to enjoy it? In some cases, perhaps we should even ask if it is ecologically appropriate or sensible to build on the coast. In some fragile coastal settings, is the price of living in paradise too great if we inadvertently destroy what we seek to experience?

Sea-Level Rise in Deep Time

Sea level rises and sea level falls. It's all a matter of time frame. During the last great ice age, some twenty thousand years ago, sea level was 400 feet lower than today. What is beachfront property now, from Tampa to Key West to Miami, was then stranded high and very dry. The northern half of North America was buried under a mountain of ice about two and a half miles thick. Nature's deep freeze had locked up huge amounts of water in this gigantic continental glacier. Then, about ten thousand years ago, a radical change happened. The global climate flipped from a frigid, "ice-age" mode to today's relatively warm and ice-free mode. Earth's postglacial temperature rose on average some 9°F, to the modern-day, overall global value of about 60°F. The northern ice sheets melted away, retreating toward polar regions. Torrents of glacial meltwaters flowed back to the ocean basins, and sea level responded by flooding exposed coastal zones. Only in the last several thousand years has the ocean's surface approached its modern-day position. Changes are still taking place. We know, from water-surface measurements taken at different locations, that sea level has risen as much as 4 to 8 inches in just the past century, in response to an increase of 1°F in average global temperature.[2]

Sea-Level Rise in a Greenhouse World

The long view of ancient history is nice as a context, but what good will this kind of hindsight do? How is the past important to us right now? Tom

Wigley and S. C. B. Raper, both formerly with the Climatic Research Unit at the British University of East Anglia, use their knowledge of events in deep time to make the best forecasts currently available for sea-level rise in a Greenhouse world.[3] Working with the ultimate scientific authority for Greenhouse issues, the Intergovernmental Panel on Climate Change (IPCC), Wigley and Raper project the changes we can expect in the near future for global sea-level position (figure 4-1). Their "best guess" factors in current political initiatives and international protocol agreements to reduce future emissions of carbon dioxide, CFC gases, and aerosol particulates. Their BAU (business as usual) scenario is shown on figure 4-1 by the bold line cutting through the middle range of possibilities. This predicted BAU scenario presumes no new policies to mitigate Greenhouse warming.

By the Boomer Breakpoint year 2025, Earth's average temperature will rise by 1.3°F, and global sea level will rise 5 inches. This accelerated change

Figure 4-1. **IPCC Greenhouse-world projections for sea-level rise.** Six scenarios show global sea-level rising in response to rising amounts of carbon dioxide in the atmosphere (fig. 3-1) and especially rising global temperatures (fig. 3-2). (Modified from Wigley and Raper, "Global Mean Temperature and Sea Level Projections under the 1992 IPCC Emissions Scenarios," see fig. 3-2)

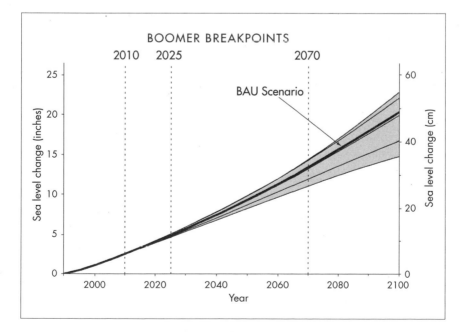

in sea level *over the next 25 years* is as much as has taken place *during the last 100 years*. By the year 2100, a global temperature increase of another 5°F should cause an increase in sea-level position by twenty inches, four times the historic rate of sea-level rise of the last century.

A recent study by the U.S. Environmental Protection Agency (EPA) spells out the Greenhouse-world consequences: By the year 2100, global sea level will rise by 1½ feet. The EPA cautions, however, that this may be a conservative estimate. There is a 1 percent chance that this rise will exceed 3 feet above 1990 levels.[4]

These projections of rising global sea level can be tailored more realistically to specific communities along the U.S. coastline, to your chosen stretch of seaside (see the maps for A.D. 2010 and 2030 in the PEW report, available online, cited in note 1). Some portions of the Pacific coastline, such as from Oregon to Alaska, are actually being geologically uplifted faster than sea level rises. Cliffs along much of California's shoreline will be uplifted at about the same rate as rising ocean waters; only several hundred square miles will be flooded out, and those will be concentrated primarily in low-lying areas near coastal cities like San Francisco.[5] Other coastal areas, however, particularly along the Atlantic, are sinking into the sea, becoming submersed because of both natural and human-caused processes. In a Greenhouse world, shorelines may erode at faster rates. Salt marshes will be displaced by the advancing brackish waters of marine embayments and estuarine sounds. Natural settlings of the land surface, such as in coastal Louisiana,[6] may be accelerated with the building of protective levees and dikes, which will shunt sediment supply away from beaches. Even the land subsidence caused by the pumping of subterranean resources, such as fresh drinking water or petroleum, will play a role.

James Titus and Vijay Narayanan, in their report *The Probability of Sea Level Rise*, offer an intriguing and reasonable possibility.[7] Using their formulas, you can calculate a local sea-level curve for your own coastal community, or for one where you may be thinking of someday purchasing a retirement house. Each location has a unique history of land uplift or subsidence relative to the ocean surface. The historical rate of global sea-level rise has been about ½₀ inch per year along the California coast, for example, more or less keeping pace with observed rates of geological uplift—resulting in apparently stationary or slowly subsiding shorelines. In contrast, the ocean is rising twice as fast as the global average along the Atlantic Coast, and along the central and western Gulf Coast it is rising five to twelve times faster than that!

Let's take a specific example and calculate how future sea-level positions along Chesapeake Bay will impact the metropolitan area of Washington, D.C. The historic rate of overall sea-level rise is ⅛ inch per year, according to Titus and Narayanan. For the 35-year interval from 1990 to 2025, local sea level will rise almost 4½ inches. *on top of* the global increase of 2 inches. By the Boomer Breakpoint year 2025 (combining both local and global components), the sea level therefore will be about 6½ inches higher than it stood in 1990. For the longer term, best estimates of sea-level elevations are 1 foot above 1990 levels in 2050, rising to 1½ feet by 2075, 2 feet by 2100, over 3 feet in 2150, and as much as 4½ feet higher than present levels by 2200. This magnitude of vertical rise will be enough to inundate a large sector of our nation's capital.

What is behind global sea-level rise?

The accelerating rise of global sea level is tied to the expanding volume of both warming marine water and freshwater quantities supplied by melting glaciers.[8] Warming air temperatures, particularly over land at high elevations and high latitudes, shift the balance between moisture being supplied to land-based glaciers and water being released from the melting ice. Mountain glaciers such as those in the European Alps are sensitive to even subtle climatic changes. If the earth's mountain glaciers melted away completely, global sea level would rise 1 to 2 feet. Today, the world's two remaining great ice sheets, the Greenland and Antarctic Ice Caps, release freshwater to the ocean through summer runoff of meltwater. They also produce icebergs as large blocks of ice break off into marine waters along the margins of ice shelves. Should the Greenland ice cap melt away completely, sea level would rise about 25 feet. Total meltdown of the Antarctic ice sheet would contribute another 200-foot rise. Fortunately, the great polar ice caps are not as vulnerable to Greenhouse warming as are smaller mountain glaciers.

In the near future, however, the modest rise in projected sea level will be driven primarily by volumetric expansion. Seawater expands in volume as it warms and as it becomes less salty and therefore less dense. The density of seawater also decreases as the freshwater runoff contributed by melting glaciers and ice shelves dilutes it near the surface. When Greenhouse-warmed and diluted surface waters then mix with deeper marine waters, thermal expansion of this entire water mass increases the overall ocean volume and thus raises sea level. As a result of all this, many seaside communities are gradually being flooded.

The greater threat

Rising sea levels are not in themselves the most critical environmental problem posed by oceans along the coastal zone. What's going to get us Boomers in the Greenhouse world is not sea-level rise—it's hurricanes and their accompanying storm surges! The problem lies in the expanding geographic area in which sea-surface temperatures exceed 80°F. This temperature is the *critical lower limit* for spawning major tropical storms.

Tropical storms are classified based on their intensity (the drop in barometric pressure at their center) and the maximum sustained speeds of winds rotating around the "eye" of the storm. The now widely used **Saffir-Simpson scale of tropical storm intensity** was developed by Herbert Saffir, a consulting engineer, and Robert Simpson, former director of the U.S. National Hurricane Center.[9] The Saffir-Simpson scale initially used critical thresholds of maximum wind speeds. Later, measured ranges of atmospheric pressure were added by Neil Frank, also a director of the National Hurricane Center.

Table 4-1 shows the threshold values for low pressure and wind speed, along with typical heights of coastal storm surges. The U.S. National Weather Service uses these guidelines in order to monitor the peril posed by hurricanes to North American and Caribbean coastlines. Hurricanes of category 3 or greater are considered "major" or "intense" destructive hurricanes.[10] During the last century, only 1 in every 100 hurricanes that directly struck the U.S. mainland was of the catastrophic category 5. Ten in every 100 were extreme category 4, 30 were extensive category 3 hurricanes, 23 were in the moderate category 2, and the remaining 36 were in the lowest, minimal category 1.[11] This means that over one-third of all hurricanes on record were severe weather events, ranked as category 3 or greater. Historically, the hurricane season runs from early June through November, with peak occurrence in early September.[12]

Type of Storm	Hurricane Categroy	Low Air Pressure at Storm Center (inches)	Low Air Pressure at Storm Center (millibars, mb)	Maximum Wind Speed (knots)	Maximum Wind Speed (mph)	Height of Storm Surge (feet)	Coastal Damage
Tropical Depression or Cyclone	TD	—	—	<34	<39	—	—
Tropical Storm	TS	—	—	34-63	39-73	<4	—
Hurricane	1	>28.9	>980	64-82	74-95	4-6	Minimal
Hurricane	2	28.5-28.9	965-980	83-95	96-110	6-9	Moderate
Hurricane	3	27.9-28.5	945-965	96-112	111-130	9-13	Extensive
Hurricane	4	27.2-27.9	920-945	113-134	131-155	13-18	Extreme
Hurricane	5	<27.2	<920	>134	>155	>18	Catastrophic

Table 4-1. **The Saffir-Simpson scale of hurricane intensity.** (Adapted from the National Hurricane Center/Tropical Prediction Center, National Weather Service Miami Forecast Office, Florida International Univeristy, Miami, Florida.)

Surf's up: Storm surge

The Saffir-Simpson scale provides a useful but general guide to the potential power of hurricanes; atmospheric pressure and wind speeds are the best gauge for predicting total wind impact. The actual damage to coastal real estate may depend, however, far more on the magnitude of storm surge rather than winds. Storm surge stacks on top of the daily tidal range, which makes it more significant in some regions than in others. For example, tidal range is about 1½ feet along the northern coast of the Gulf of Mexico, predictably focusing the surge flooding in a narrow vertical range. But tides may rise and fall over a range of 9 feet along the southern Atlantic Seaboard, depending upon the time of year—potentially magnifying at high tide or even canceling out storm surge flooding at low tide. Of course, as hurricanes track inland from the coastal zone and perhaps stall out as downgraded tropical depressions, the combination of strong winds, torrential rainfall, and slowly moving storms heightens the danger from flooding as well as from wind-thrown trees falling across utility wires, homes, and cars.

The Saffir-Simpson estimates for storm surge are much less secure than for hurricane winds, but they are the basis for evaluating a storm's potential for property damage. When specific hurricanes threaten our coast, regional hurricane planners for FEMA (Federal Emergency Management Agency) use precise data on the known storm track and speed, projected timing of landfall relative to times for high tide, and even the slope and water depth of the ocean floor adjacent the shore. Their computer model (called SPLASH) predicts the spatial extent of storm-generated surges that local officials use to coordinate emergency evacuations before hurricane landfall.

The National Hurricane Center tells us that for a category 1 hurricane one should plan for winds of 74 to 95 mph, with surge levels typically 4 to 6 feet above normal, causing minimal damage.[13] Wave attack will be focused on surfside piers, with limited flooding of beachfront homes and roads. Winds will knock over poorly constructed signs and mobile homes not properly anchored to their foundations.

A category 2 hurricane, with higher winds reaching up to 110 mph and driving inland surges to 6 to 9 feet above normal, will cause moderate structural damage to mobile homes as well as to oceanfront piers and small watercraft in unprotected moorings. Winds will blow down some trees, prune their limbs, rearrange roofing shingles, and damage windows and doors of homes that are not boarded up.

The fierce winds (111 to 130 mph) and major storm surges (9 to 13 feet above normal) of a destructive category 3 hurricane will extensively destroy mobile homes, rip open roofs and walls of small homes and utility buildings, shred foliage from forest canopies, and cause windthrows of large old trees. Coastal lowlands up to 5 feet in elevation will be inundated; low-lying homes will be flooded and battered by wave surf and by floating debris. Evacuation may be required for coastal homes within several blocks of the shoreline.

Extreme hurricanes of category 4 will flood terrain up to 10 feet in elevation requiring massive evacuation of residential areas as far inland as six miles. Their very high winds (up to 155 mph) and surge attack (generally 13 to 18 feet above normal ocean-surface levels) will cause major beach erosion, massively damage maritime-forest trees, eliminate mobile homes, and ravage lower floors of multistoried homes and businesses; some roofs may collapse. The nation's most rigorous building code, the South Florida Building Code for Dade County, requires that new construction be built to withstand wind speeds up to 140 mph.[14]

Catastrophic hurricanes of Saffir-Simpson category 5 have sustained winds greater than 155 mph and storm surges more than 18 feet above normal. One can expect complete destruction of mobile homes and roof collapse and building failure for residences, with massive damage to lower floors of all structures within 500 feet of the ocean shore and with elevations less than 15 feet above mean sea level. Water damage inside homes will be also caused by extensive shattering of residential windows and doors.

Observations of recent extraordinarily intense hurricanes have led meteorologists to ponder the need for an even higher level of classification, which would be category 6 on the Saffir-Simpson scale. There exists the distinct possibility that such hurricanes of unprecedented magnitude will occur in a Greenhouse world.

How Will a Greenhouse World Affect Hurricanes?

Storms will become more frequent as well as more powerful in the Greenhouse world. Tropical depressions (cyclones), tropical storms, and hurricanes all form over marine waters between 5 and 35 degrees latitude north of the equator. This sector includes equatorial waters of the North Atlantic Ocean, the Gulf of Mexico, and the eastern North Pacific Ocean.[15] Evaporation from warm ocean waters provides the energy for tropical storms to develop, then intensify (it releases the latent heat of water with the physical transformation to water vapor). Evaporation rates increase exponentially with rising surface-water temperatures. To sustain hurricane development, sea-surface temperatures must be above the critical threshold of 80°F. When ocean surfaces drop below this temperature—for example, during wintertime in the North Atlantic—it is too cold for tropical storms to form.

Raper notes that we should expect "a marked change in the frequency of tropical storms" by the year 2050. His studies show that above the threshold of 80°F, each increase in sea-surface temperature of about 2°F produces a 41 percent increase in major tropical storms in the North Atlantic Ocean and a 38 percent increase in the eastern North Pacific Ocean.[16]

In the North Atlantic, where three to eight hurricanes currently develop in a typical year, the overall frequency of hurricanes has increased over the past century as ocean water has warmed. Figure 4-2 shows three seasonal

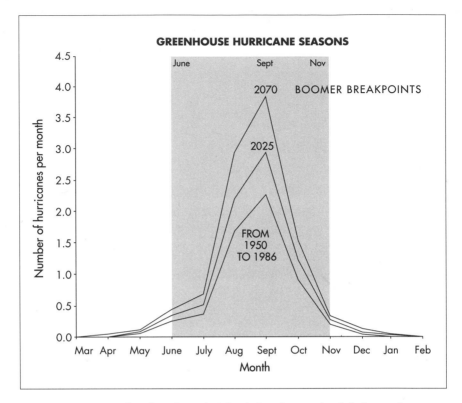

Figure 4-2. **Seasonal cycles of North Atlantic hurricanes.** As global warming accelerates, warmer ocean spawning grounds will produce more hurricanes, and their peak season will expand. (Modified from S. C. B. Raper, "Observational Data on the Relationships between Climate Change and the Frequency and Magnitude of Severe Tropical Storms," pages 192–212, in R. A. Warrick, E. M. Barrow, and T. M. L. Wigley, eds., *Climate and Sea Level Change: Observations, Projections, and Implications* [Cambridge: Cambridge University Press, 1993].)

curves for present-day and Greenhouse-world hurricanes. The lowest curve shows modern hurricane frequency, with a prominent peak in September. The intermediate curve shows the seasonal hurricane cycle we project for the Boomer Breakpoint year 2025, presuming a 1.3°F increase in mean global temperature and 29 percent increase in storm frequency. The third curve shows the seasonal pattern we calculated for the Boomer Breakpoint year 2070, consistent with a 3.0°F increase in mean global temperature (figure 3-2) and a 70 percent increase in storm generation.

We view these scenarios for Greenhouse-world hurricanes as *conservative*. Near-future sea-surface temperatures for the North Atlantic should *substantially exceed* the modest increase of mean global temperatures we used in these projections. Nevertheless, our conservative input yields stunning results on figure 4-2. Today's average of two September hurricanes should increase to at least three by 2025, and to four or more hurricanes by the year 2070. The autumn maximum in hurricane development may extend into early October, and the hurricane season may start as early as April and end as late as December. In a Greenhouse world, warm marine water (greater than 80°F) will cover a larger geographic area for a longer period of each year. The temperature contrast will be heightened between land and ocean. The ocean will be slower to cool below the critical threshold of 80°F in the wintertime. These Greenhouse factors will combine to increase the duration of the hurricane season, with more hurricanes produced more frequently from a larger tropical spawning area.

If the North Atlantic will produce more storms, what about the *tropical* Atlantic? Researchers Leonard Druyan and Patrick Lonergan have used the GISS GCM model (with doubled levels of carbon dioxide) to spell out the ramifications of global warming for tropical storm development by the year 2070.[17] As storms form in the Cape Verde region off the West African coast, their winds will tend to spin faster, and a greater proportion of storms will strengthen into full-blown hurricanes there as well.

In the twenty-first century, hurricanes are also likely to be more intense, with lower atmospheric pressure achieved at their centers. Futuristic computer simulations, based on a sophisticated version of the GISS model that links atmospheric and oceanic circulation patterns, project that August sea-surface temperatures in tropical oceans will rise 4 to 9°F over today's values. (According to model results simulating Greenhouse-world weather patterns, even relatively small increases in sea-surface temperatures cause major increases in hurricane intensity. For example, an increase of 5°F can mean a 20 percent increase in maximum sustainable wind speed and up to 40 percent intensification of low pressure within the eye of the hurricane.) According to climate modeler K. A. Emanuel, "Were these sea surface temperature increases realized, the maximum destructive potential of tropical cyclones would be substantially increased, in some places by as much as 60 percent."[18] The hurricanes of greatest magnitude (category 5 *or greater*) would form most predictably in partly enclosed basins such as in the northern Gulf of Mexico and in the Gulf of California near the Baja Peninsula of Mexico.

William Michener and colleagues have described the ecological responses to near-future Greenhouse environmental changes that we should anticipate for coastal wetlands of the southeastern United States.[19] They document the geographic distribution of direct coastal hits by hurricanes over the ninety-four-year interval from 1899 through 1992.[20] A total of 219 historic hurricane landfalls were reported along the Atlantic Coast from Maine south to Florida and along the Gulf Coast from Florida west to Texas. Of these, 89 were major hurricanes of category 3 or greater. These historic hurricane data show that, for example, the Texas shoreline is typically struck by a hurricane every third year, although it usually has been six years between impacts by major hurricanes of categories 3, 4, or 5. In contrast, the neighboring state of Louisiana generally receives the full brunt of a hurricane every four years and a major storm every eight years. Florida, with coastlines bordering both the Gulf of Mexico and the Atlantic, was hit 55 times by historic hurricanes, including 23 landfalls by major hurricanes. This historic record translates into an average return time of two years between hurricanes and four years between hits by category 3 or greater hurricanes.

Farther north along the Atlantic Coast, the protruding headlands of the Carolinas and of southern New England sit squarely in the track of many northward-moving storms, with hurricane return times occurring about every five and fifteen years, respectively. But two zones are noticeably protected from the onslaught of most hurricanes. Coastal Georgia, with a record of only five historic landfalls over the ninety-four-year record, has a typical recurrence interval of nineteen years between hurricane landfalls and greater than ninety-four years for category 3 events.[21] Why is this? Starting from eastern Florida, the Atlantic coastline bends back to the west along southern Georgia. The Gulf Stream moves warm, tropical water northward along Florida's eastern coast, arcs up to fifty miles offshore from Georgia, and then once again returns toward the Carolina shore. It is finally deflected eastward by the obstructing South and North Carolina strand of barrier islands. The contrast in sea-surface temperatures marking the western edge of the Gulf Stream acts like a powerful magnet—attracting, capturing, then deflecting northward most wayward hurricanes that might otherwise strike shore in the Georgia embayment. So the Gulf Stream offers "favored status," providing a *protective hurricane shield* for coastal Georgia.

The second sheltered zone occurs northward of the Cape Hatteras shield, in the scallop of protected shoreline extending from Virginia to New Jersey. The combination of a protected location and generally colder ocean waters (below the 80°F threshold) most of the year markedly reduces the vulnerability of this area to the threat of hurricanes at present.

How will the hurricane threat change in our lifetimes?

Today, along both the Gulf and the Atlantic Coasts, one or two hurricanes typically strike someplace each year, and a category 3 to 5 hurricane every other year. By contrast, for the Boomer Breakpoint year 2025, we forecast an annual average of two hurricane landfalls somewhere along the Gulf Coast, and three such events routinely by the Boomer Breakpoint year 2070. The incidence of major hurricanes along each coast will accelerate to at least one extreme impact each hurricane season. By 2070, the resort Gold Coasts of Florida and Texas should expect hurricane landfalls that are twice as frequent as today's, with category 3 or greater hurricanes striking Florida every other year and making landfall in Texas every third or fourth year.

The PEW report on Greenhouse-world impacts along coastal America spells out the magnitude of personal property risk from hurricanes. Between 1988 and 1993, the total value of insured property rose 69 percent, reaching $3.15 trillion in coastal counties from Texas to Maine. Over a longer period, from 1960 to 1995, Florida's population tripled to 14 million. Hurricane Andrew, which struck southern Florida in 1992, caused $30 billion in damages, the costliest storm in U.S. history. As retiring Boomers continue to settle along the Florida shoreline, they may be underestimating the storm threat to human life and property. Attracted to expensive beachfront homes and high-rise condominiums, many relocating Boomers place themselves and their families directly in the paths of more frequent and more powerful hurricanes. Using past storms for precedent, future hurricane costs to property should average $5 billion per year.[22]

Internationally recognized climatologist Tom Wigley, of the National Center for Atmospheric Research (NCAR) in Boulder, Colorado, cautions that Greenhouse-world hurricanes will typically carry much more energetic punch than those of the twentieth century, with far stronger winds and substantially more rainfall than delivered previously.[23] Remember the lesson your insurance company knows all too well: In a hurricane, much of the water damage to a given home is caused by the combination of strong wind, shattering windows or breaching roofs, and wind-driven torrential rains, *not* by flooding. This kind of rain-related water damage, however, is routinely *excluded* from the standard version of the homeowners' comprehensive policy. Open to suggestions? If you are determined to own a permanent residence in a vulnerable coastal zone, and you are willing to take the risk of being hit by Greenhouse-world hurricanes, one way to protect yourself and your belongings from total financial loss is to fork over the extra cash for a water-damage rider on your insurance policy.

But what is a vulnerable area? In the Greenhouse world, how far north along the Atlantic Coast will the zone of frequently recurring hurricanes extend? Charles Coutant of Tennessee's Oak Ridge National Laboratory has examined latitudinal shifts in suitable habitat for marine fish along the eastern coast of North America.[24] His data apply just as well to suituable "habitat" for hurricanes. Coutant presents the typical range in monthly temperatures for offshore sea-surface waters from northeastern Florida to northern Maine and Nova Scotia (figure 4-3). Historically, only marine water bordering Florida, Georgia, and South Carolina has reached temperatures of 80°F or greater during the interval from June to September. This corresponds with both the temperature threshold necessary for hurricane development as well as the modern zone of frequent hurricane landfalls from East Florida to the Carolinas. Coutant has also used the results from two Greenhouse-world climate models to simulate latitudinal and seasonal temperatures along the Atlantic Coast for the year 2070, assuming a doubling of atmospheric carbon dioxide from preindustrial levels. Both computer simulations forecast elevated sea-surface temperatures throughout all seasons of each year. Both also indicate that the Gulf Stream will weaken off the Atlantic Coast as the temperature contrast diminishes across the western margin of this warm ocean current. This oceanographic change should markedly reduce the Gulf Stream's effectiveness in steering hurricanes offshore and away from the southern Atlantic Coast.

The best-case scenario of least change for Boomer Breakpoint 2070 is projected by the GISS model (figure 4-3). In this projection, the stretch of coast potentially influenced by most frequent hurricanes expands north-

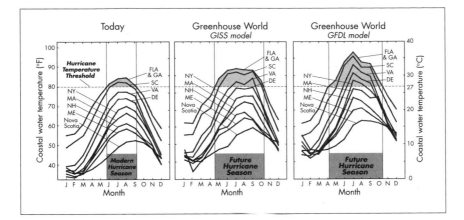

Figure 4-3. **Hurricane zones along the Atlantic seaboard.** According to both the "best-case" (GISS) and the "worst-case" (GFDL) models of global climate, hurricanes will be able to strike the coast farther and farther north as climate warms. The shaded zone above 80°F identifies the potential hurricane season along the entire seaboard, from northeastern Florida to Nova Scotia. (Modified from Charles C. Coutant, "Temperature-Oxygen Habitat for Freshwater and Coastal Striped Bass in a Changing Climate," *Transactions of the American Fisheries Society* 119 (1990): 240–253.)

ward to include Virginia, with substantial storm activity anticipated from late April through late October. The worst-case scenario of greatest change for Boomer Breakpoint 2070 is simulated by the GFDL computer model (figure 4-3). According to the more extreme GFDL model, the primary zone for hurricane impact will stretch from Florida to New York, with the hurricane season spanning from April through at least late October. The bottom line is that, in a Greenhouse world, the full Atlantic seaboard from the Florida Peninsula to southern New England should receive both a greater number of hurricanes and more intense hurricanes over a longer season. Even the two coastal areas that are relatively safe sites today, along eastern Georgia and from Virginia to New Jersey, will be increasingly prone to severe storm damage.

The four-bite rule

Our insurance agent speaks of the two-bite rule in insurance coverage. He tells us, "If your dog bites your neighbor, that's my problem. Homeowners' insurance will cover your neighbor's medical expenses. However, if your dog bites your neighbor a second time, then that's *your problem!* We won't cover you when he sues."

FEMA, the Federal Emergency Management Agency, is working on something akin to a **four-bite rule.** FEMA is empowered under a series of congressional acts to offer private citizens voluntary flood insurance within coastal zones and river floodplains.[25] In order for citizens to qualify for this insurance, their local communities must agree to participate in the **National Flood Insurance Program (NFIP),** must comply with the state floodplain management, specifically the Coastal Barrier Resource Act as commonly overseen by the state's Department of Natural Resources (DNR), and must implement and regulate new construction based on existing building codes.

For each participating community, the United States Army Corps of Engineers has designated **Special Flood Hazard Areas** on specialized elevational maps called **FIRMs (Flood Insurance Rate Map)** or **FHBMs (Flood Hazard Boundary Map).** These Special Flood Hazard Areas are defined as having a 1 percent chance of being flooded in a given year. NFIP officials are careful to state that this does not represent the flooding risk associated with an extreme, dreaded "once-in-a-hundred-years" flood. Unfortunately, such rare and extreme events will become more commonplace in the Greenhouse world. Today's hundred-year hurricane will occur every seventy years by Boomer Breakpoint year 2025 and every sixty years by Boomer Breakpoint 2070. Correspondingly, the federal mandate of 1 percent probability per year translates into ever broader flood-impact zones generated by ever more powerful hurricane landfalls. FEMA's Flood Maps will have to be continually revised, constantly increasing the cost of flood insurance.[26]

Under current rulings, FEMA offers up to $250,000 for each residential structure and another $100,000 for contents of single-family homes. For condominium complexes, the condo homeowners' association may apply for comparable coverage of $250,000 per residential unit. Private market companies may offer co-insurance for additional coverage beyond the NFIP limits. *But buyer beware.* In some regions, *no* private insurance companies are willing to offer supplementary coverage to pre-FIRM houses, built before the standard FIRM regulations were first applied to a particular area. As a result, your coverage may be capped to the extent of FEMA's subsidy of your policy.

NFIP insurance rates are determined by the construction date of the home, its location on the Flood Insurance Rate Map (FIRM), its proximity to the ocean beachfront, salt marshes, and tidal creeks, and its compliance with building code ordinances. More critically, NFIP insurance rates are tied to the elevation of the home's first floor (a base-level elevation for 1 percent flood probability per year, typically 13 to 19 feet above modern mean sea level). NFIP offers hurricane flood insurance to older homes under the grandfather clause—homes that predate the enactment of the coastal laws or the publication date of the local FIRM maps. However, following substantial damage (defined as more than 50 percent loss) or complete destruction of the home by hurricane flooding, individual states may require an additional setback distance from the ocean shore for new construction.

So what *is* the **four-bite rule?** NFIP is developing new legislation to target repeat offenders, that is, the relatively few homes that are repeatedly flooded out, then repaired or rebuilt with federal flood insurance. This new NFIP class of problem homes, with a troubled track record of multiple flood losses, is called *target repetitive loss buildings*. They are characterized as a "building with four or more losses, or with two or more flood losses cumulatively greater than the building's value." Targeting this suite of roughly ten thousand problem homes should dramatically reduce the $200 million in flood-damage claims currently paid out each year. NFIP's repetitive law strategy is proposing corrective mitigation measures for these problem homes that are flood damaged four or more times. The federal government may voluntarily offer to purchase the problem home outright. Alternatively, mitigation efforts may require it to be relocated to a more flood-protected site, elevated above base flood level, or simply repaired to "flood proof" the residence before a new insurance policy will be made available. As of May 1, 2000, all requests for insurance renewal require the owner's compliance for mitigation *before* these repeat problem homes are offered a new NFIP policy.[27]

Availability of flood insurance is a powerful factor influencing a bank's willingness to provide a construction loan or long-term mortgage. In the future, more homes will fall under the four-bite rule as coastal areas become increasingly vulnerable to storm surge. The wary home-buyer will check the relevant FIRM map in advance of placing earnest money down on a coastal residence, to make sure that the site is habitable now *and* in the future Greenhouse world.

Deal with It, Move It, or Lose It

In the face of massive, rapid environmental change, nature's solutions have been straightforward: (1) *deal with it* (adapt); (2) *move it* (migrate to a more suitable environment); or (3) *lose it* (go extinct). What are some of the solutions being considered for coastal zones in view of the Greenhouse threat of increasingly severe flooding? How can shoreline communities prepare for sea-level rise? Is this a situation that lends itself to a simple technology fix? Can the U.S. Army Corps of Engineers just build higher protective seawalls?

James Titus, of the U.S. Environmental Protection Agency (EPA), offers a suite of adaptive responses to prepare for rising sea level (figure 4-4).[28] His first premise is that *no action is no solution*. Without protection, the developed sectors of barrier islands and the ocean shore will be eliminated during extreme storm events. It's just a matter of time. Given a legislative mandate and appropriate levels of support, he argues, the Corps of Engineers can build protective structures such as higher bulkheads, seawalls, and rock revetments to hold back the sea. Levees constructed around entire developed barrier islands can provide one kind of short-term solution. As sea level continues to rise, however, communities will slip below flood level. That can be dealt with as well. New Orleans, which is already six feet *below* sea level, is fortified behind more than one hundred miles of barrier walls atop levee ridges, and it has vintage pumping stations (built circa 1915) to remove rainwater and groundwater collecting within depressions behind the river levees. Other adaptive responses include stronger building codes for better "flood proof" and "wind proof" residences exposed to successively stronger storms.

Another alternative for coastal communities is to move it—move up or move back! In a piecemeal fashion, local zoning boards can "raise the bar" by literally raising houses on stilts. The zoning solution may be to raise the minimum elevation of the home's first floor, with the hope of leading rather than lagging behind rising sea level. Instead of raising each house piecemeal, another possibility is to raise the surface of the island (figure 4-4). This solution has been used in Galveston, Texas, where the land is subsiding and sediment fill has been used to offset sea-level rise. After the catastrophic

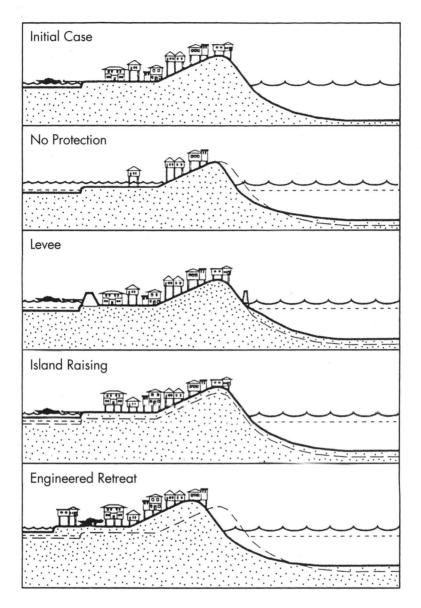

Figure 4-4. **Protective responses to sea-level rise.** Given the value of shore-line properties, "no protection" is not a likely option. (Modified from D. S. Shriner, R. B. Street, et al., "North America," pages 253–330, in R. T. Watson, M. C. Zinyowera, and R. H. Moss, eds., *The Regional Impacts of Climate Change: An Assessment of Vulnerability, A Special Report of IPCC Working Group II* [Cambridge: Cambridge University Press, 1998].)

hurricane of September 1900, people of the city of Galveston constructed a concrete seawall more than three miles long and seventeen feet high. They dumped earthen fill behind the barrier to raise the ground surface by seventeen feet in elevation, and they raised 2,156 buildings on pillars above the new grade.[29] In other coastal cities like Miami, Florida, using clayey landfill to elevate the land surface can provide a cap to seal off the underlying bedrock of porous limestone. The clay seal stops groundwater from leaking upward and flooding the community.[30]

"Move back" is another mantra for the coastal land planner. In theory, this may be achieved today by using either setbacks or engineered retreats. Setbacks are zoning regulations that block new oceanfront construction; it must be set back from the shoreline in areas vulnerable to wave-attack erosion. But this potential infringement of personal rights was challenged successfully in the U.S. Supreme Court case of Lucas versus the South Carolina Coastal Council.[31] In contrast, the legal right of property owners to develop their own land can be protected with rolling easements, innovative solutions now used in Texas and South Carolina. Rolling easements represent the adaptive strategy of engineered retreat from rising sea level (figure 4-4). Easements for new coastal construction are permitted if property owners develop their land such that ocean beaches and wetlands will continue to shift naturally, to roll on inland with time. This limited-time easement for the homeowners remains valid until the Greenhouse-world realities of storm surges and rising sea level reclaim their home site. Under this strategy, ocean beaches, defined as the land exposed between low and high tides, remain under public domain. Rolling easements on the beach strand thus counterbalance the extent to which new construction in burgeoning communities can encroach upon or diminish this public space.

Beachfront Homes: Paradise for Years to Come?

How will storm surges produced from recurring Greenhouse hurricanes affect oceanfront real estate? Figure 4-5 presents the best available forecast for the mean position of global sea level through the third Boomer Breakpoint in the year 2070. In the baseline year, 1990, the typical storm surge during landfall of a category 4 hurricane flooded coastal zones with land elevations as high as 13 to 18 feet above mean sea level (see table 4-1). Emanuel predicts a 60 percent increase in hurricane magnitude (and pre-

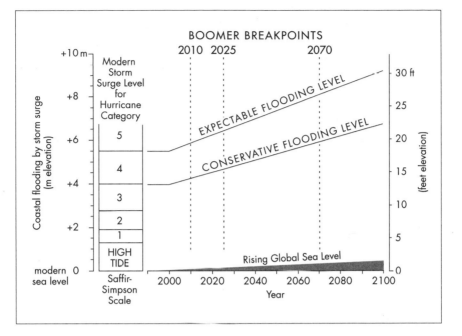

Figure 4-5. **Hurricane storm surge and sea-level rise in a Greenhouse world.** Boomers planning a seaside retirement can use this chart to find relatively safe building or purchase sites above storm surges from all but the strongest, category 5 hurricanes—which unfortunately will become more common in the Greenhouse world.

sumably storm surge) by the year 2070. We use this as a basis to calculate the progressively greater heights of coastal land that will be flooded in a Greenhouse world.[32]

FEMA's primary criterion for offering voluntary flood insurance is the **Base Flood Elevation,** which identifies coastal sites that have a 1 percent chance of being flooded in a given year. FEMA will reset and raise this Base Flood Elevation in response both to rising sea level and to the heightened flood levels of hurricane storm surges.

In choosing the specific location for your oceanfront retirement home, you need to consider several significant issues: risk level, site elevation, and the relevant time window of opportunity—when to get in and especially out. Figure 4-5 shows two curves plotted for the twenty-first century, identifying modern-day elevations of beach frontage most likely to be flooded by the Greenhouse-world combination of sea-level rise and storm surge predicted for major hurricanes. The lower curve conservatively indicates the coastal zone elevation that will be inundated every few years in this Greenhouse world. Even if you *enjoy* being flooded, the four-bite rule works against you there. The upper curve is the scenario you can use to *plan* your building decisions. This upper curve indicates the probable flood level for relatively infrequent but expectable catastrophic hurricanes, which will strike about every twenty-five years. Your personal strategy should be tied to your retirement or moving time, your tolerance to level of risk, and long-term plans for deeding your home to your children or grandchildren in the twenty-first century.

Figure 4-6 illustrates the Greenhouse-world reality for one such retirement investment: a resort development on a prestigious barrier island. The first step is to evaluate the *investment lifetime* of your prospective home with regard to the eroding shoreline and rising flood level. First, your local zoning board and state department of natural resources have established local rates of shoreline retreat based on historic survey maps dating back to the 1860s as well as aerial photographs from the 1930s. For example, historic coastal zones of erosion and accretion are mapped for many of the barrier Sea Islands of South Carolina and Georgia.[33] You can use these **Coastal Resource Atlases** of shifting shorelines to predict your personal setbacks, or safety zones, to forecast how far beach frontage will erode by the Boomer Breakpoint years 2010, 2025, or 2070. Remember, FEMA doesn't insure property in the coastal "zone of imminent collapse" adjacent to the (rising) marine limit of mean high tide.

The second step is to pick a site elevation tied to your time frame and level of risk (figure 4-5). If, for your lifetime, you want a high and dry home, choose a land-surface elevation above 22 feet for the year 2025, and above 27 feet for the year 2070. If the home is the legacy intended for your grandchildren, you may wish to choose an even higher site, above today's elevation of 30 feet (figure 4-6).

In researching the environmental setting for an area that particularly interests you, you may find that the risks greatly outweigh the potential benefits of being at the coast. Some barrier islands, no matter how enticing they may seem in advertising brochures, may be just too exposed to hurricane damage or may contain ecosystems too fragile to withstand the pressures a

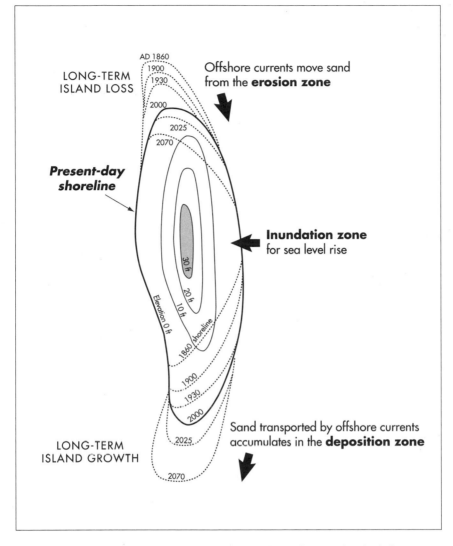

Figure 4-6. **Barrier island erosion, accretion, and inundation.** The shaded area marks "paradise found"—the optimal site for your island bungalow, perched high and dry above the rising Greenhouse-world seas and storm surge. The dashed positions of moving shorelines portray long-term trends of island loss and accretion, as offshore currents redistribute beach sand and as shorelines shift inland with the ever rising sea level. The contour lines of 0, 10, 20, and 30 feet show island elevation above *today's* sea level.

resort development will entail. In such a case, it may be best to reconsider your options. But in any case, living well means not only getting to the coast and building a dream house. It also means becoming a steward of the land. This may require setting aside vulnerable ecological areas in nature preserves, which may also serve as aesthetically pleasing adjuncts to human communities.

The *worth* of coastal zones as spectacular places for retirement homes is great; the *value* of undisturbed stretches of beach as sanctuaries for sea turtles and migrating birds is inestimable. Along the Gulf Coast of the Florida Panhandle, developers seek to accommodate both nature and resorters. An appropriate solution is to intersperse private coastal developments (set back on high bluffs) with public reserves. And along all coasts, state parks and national seashores are (or should be) located strategically to preserve barrier-island strands, nature's first line of defense against wave attack. They should preserve as well the natural sand-dune habitats, critical for maintaining diversity of both native plants and animals.

Think and plan intelligently now, with a sensitivity to the potential vulnerability of a coastal landscape to future environmental change. By planning with an eye to stewardship of the land and its biological communities, you can help ensure that your chosen coastal retirement site will be able to exist in balance with its natural setting for many generations to come. And remember, too, it may not be the most desirable location right now—as our good friend is betting with his Caribbean investment.

CHAPTER

5

The Lakeshore

Our fellow Michiganders suffered a prolonged heat wave in July of 1999, blamed in large part on La Niña, an extreme weather year following on the heels of a pronounced El Niño year in 1998. Relief from the heat was restricted to the shores of Lakes Michigan, Huron, and Superior. From the vantage point of our summer cabin on the pristine north shore of Lake Michigan, we were spectators of an abrupt exodus as our summer neighbors fled the cities "down below" the Mackinac Bridge that spans the Straits of Mackinac separating the Lower and Upper Peninsulas of Michigan. One of our neighbors arrived in the midst of this scalding regional heat wave. As he slowly eased into a lawn chair, intentionally placed half-submersed at the cooling water's edge, our friend described the unsettling scene left behind in Grand Rapids. "It was eerie, like something out of an Orson Welles movie about World War III. All the streets were deserted during the ozone alerts and the intense midday heat. Can you believe it? A hundred and five in the shade, and a heat index of a hundred fifteen degrees! It's *never* been that hot before." With the cooling breeze off Lake Michigan, a vast expanse of shimmering blue water steps away from the front door of his log cabin, our neighbor felt as if he had finally arrived at a safe site.

Today, within the Great Lakes region, most Boomers *work* within the metropolitan corridor that stretches from the Twin Cities and Madison to Chicago, Detroit, and Toronto. But many of those same people *live* for the chance to escape to their weekend and summer lakeshore cottages, which ring the margins of thousands of northern inland lakes and line the shorelines of the Great Lakes as well. Stressed-out Boomer resorters travel up to five hundred miles to find peace of mind along the lakeshore.

Check out the Web site that's fast becoming a Boomer cult "fix" for wanna-be-yoopers.[1] And if ya have to ask . . . yoopers (pronounced U-pers) are those of us who have discovered clean, fresh air, miles of nearly deserted sandy beaches to explore, endless trails through natural wilderness, and, best of all, relief from summer heat and urban overcrowding. The Great Lakes' best-kept secret lies beyond the Mackinac Bridge in Michigan's Upper Peninsula, abbreviated U.P. by the lower Michigan "trolls" who still live "down below the Bridge." Get it? U.P. resident, called a U-per? From their PC terminals, wanna-be yoopers can now watch with melancholy the continuous traffic line of cars, RVs, and campers with boats that trek north every Friday afternoon, returning south each Sunday. This is made possible by a live-action camera that scans the traffic flow on the five-mile-long Mackinac Bridge every fifteen minutes.

For some lakeside resorters, a long-term retirement strategy is to winterize a weekend bungalow and transform it into an ultimate year-round lakeshore destination. However, troubling questions about the Greenhouse world remain. These quandaries deal with key issues of personal creature comfort and recreational opportunities. How far north do we have to go to escape the summer heat of the city? Will Greenhouse warming impact our quest to fish for "big lunkers" of lake trout and coho salmon? Where can we predictably find thick snowpack for snowmobiling treks and cross-country skiing? Will a changing climate threaten those beautiful, big birch trees that now grace our rustic cabin setting? How will sport hunting for wild turkey and white-tailed deer be affected? This chapter explores these personal dilemmas, set in the context of the changing dynamics of lake ecology, snowfall patterns, and forest ecology.

How Hot Is Hot?

Global scenarios project that average temperatures will be warmer by over 1°F in 2025, by 3°F in 2070, and by 5°F in the year 2100 (figure 3-2). Based on GCM computer projections of the Greenhouse-world climate (chapter 3), on the average we expect a long-term trend toward hotter summers, longer spring and autumn seasons filled with pleasant, even balmy days, and fewer days of extreme winter cold. However, such average changes in worldwide temperatures will not be experienced everywhere and throughout each season. Instead, substantive regional departures will occur from the global "norm."

Recent studies by M. E. Schlesinger and by Tom Wigley outline both regional climate patterns and broad projections of "plausible" scenarios of Greenhouse-world climates yet to come. The results are mapped as the *difference* between Greenhouse-world conditions and those of today. Overall, given a doubling of atmospheric carbon dioxide, the GFDL model projects an average Greenhouse world increase of 7°F in air temperature and a 9 percent increase in total precipitation, relative to the present day. The simulations from the GISS model are comparable, with a warming of 7°F and 11 percent more precipitation than today.[2]

In his 1999 PEW report, Tom Wigley averaged the climate projections from fifteen different versions of GCMs—and he found that all those Greenhouse-world forecasts converge on the same conclusions. Nearly all the United States and Canada will heat up at a faster rate than the global average. The high latitudes should experience the greatest magnitude of warming. In the Great Lakes and New England regions, northern states from North Dakota and Minnesota east to Maine will warm twice as much as the global average. Greenhouse-world precipitation levels will tend to increase across most of North America, although very large differences in available moisture will persist seasonally. Deepening drought conditions should intensify during winter in the Southern High Plains of Texas and northern Mexico. During summer and fall, severe droughts will expand northward into the northern plains and western Great Lakes.[3]

Boomers will be well on their way to adapting to this new Greenhouse world by the year 2037, when the Social Security Trust Fund is expected to go bankrupt. By the Boomer Breakpoint of 2070, however, even more dramatic changes are projected. GCM results forecast warmer, drier summers and substantially warmer, wetter winters. Soil moisture may decrease by as much as 50 percent across the Great Lakes region.

Especially interesting are the model projections for rapid and major temperature transitions during spring and fall. Earlier onset of warming spring temperatures will trigger snowmelt and increase runoff to streams, wetlands, and lakes. Greater frequency and duration of "hot spells" will reduce soil moisture, depleting sources of moisture for clouds that might otherwise form over land. Less cloud cover means more sunlight absorbed directly by the ground and consequently more heat energy available to further warm the air and dry out the soil. Such a self-perpetuating drought cycle in Greenhouse-world summers may make for a midwestern dust bowl reminiscent of the 1930s. Based on current estimates, in the Great Lakes region the weather of the twenty-first century will probably be 5 to 10°F hotter than today, with longer summers. Rainfall will be patchy, although heavy precipitation events will become more frequent and more intense. Despite more total annual precipitation, accelerated year-round warming will enhance greater evaporation, dry out soils, and lead to increased frequency and severity of drought.[4]

The Resorters' Dilemma

What does a hot climate outlook mean for leisure lifestyles at the lakeshore, both in the Great Lakes region and in New England? Henry Hengeveld of the Canadian Climate Centre in Ontario has spelled out some of the implications.[5] In the near future, the delayed onset of autumn and earlier beginning of spring will mean shorter, generally warmer winters, with less frequent and less extreme intervals of very cold temperatures. Snow cover may not persist as a continuous blanket of ground insulation throughout the winter. More frequent so-called January thaws in midwinter may completely melt off snowpack in southern Ontario, the Lower Peninsula of Michigan, and southern Wisconsin. The combination of bare earth and sharp cold snaps bodes ill for residential areas, as frost heave will more frequently shatter cottage and municipal water lines buried below ground, and as freeze-thaw action will tend to deepen potholes on washboard country roads. As temperature fluctuates about the freezing point, the winter precipitation regime may alternate among episodes of snowfall, glaze-ice storms, and rainfall.

On small lakes, during the brief winter season, ice cover will decrease significantly in both thickness and persistence. This is not good news for intrepid ice fishermen and their "tip-up" communities built of portable ice-house shanties. The extended warm growing season will also pose problems for larger lakes. Hotter summer temperatures; high evaporation rates; soil moisture declines of 30 to 50 percent; greatly reduced water supply from overland runoff through smaller, ephemeral streams; and dropping groundwater tables that sap away water from lake basins rather than recharging them—all will lead to fluctuating and generally lowered lake water levels.

Hengeveld cautions that changes in the Great Lakes, as well as in the smaller inland lakes, will be far-reaching. By 2090 in the Canadian Climate Scenario, water supplies will decrease significantly, and water levels in the Great Lakes will drop by at least four to five feet.[6] Coastal marshes will disappear or recede offshore as the shoreline shifts, disrupting crucial nesting habitats for shorebirds. Annual outflow from the Great Lakes into the Saint Lawrence Seaway will diminish by some 20 to 40 percent. This diminished flow, along with lake surface waters warmed by 4.5 to 9°F, and summer blooms of aquatic algae may lead to accelerated decomposition of organic material, dropping concentrations of oxygen dissolved in surface waters, and fish kills. Lower lake levels will seriously disrupt navigation of the Sault Locks between Lakes Superior and Huron by commercial freighters, as well as of other dredged shipping channels. Winter ice cover across the Great Lakes will largely disappear, perhaps extending the ice-free navigation season for those freighters and tankers able to navigate the shallower channels between the Great Lakes.[7]

On Lake Michigan, the Greenhouse-world water level is expected to fall some two to eight feet from its historic high position, which was reached in the winter of 1986–1987.[8] This could be great news for lakeshore owners over the coming decades. At our cabin on the northern shore of Lake Michigan, the width of sandy beach grew by 40 percent with the drop in lake levels between the summers of 1999 and 2000. At this rate, our Lake Michigan beach will grow to as much as a quarter of a mile wide in coming decades. We jokingly tell our summer neighbors that we all will have to pack a picnic basket and take a day hike just to go for a swim! For some of our neighbors, however, the future change in shorelines is already a serious concern. *It now makes a difference* whether their property boundary extends across the beach and down to the average high-water line, or whether the beach belongs to all members of the subdivision, or whether it is public land. Where zoning permits, aggressive real estate developers may already be starting to plan new "bowling alley" strip corridors lined with resort homes, built outward onto the new stretch of beach created by the

Greenhouse-world lowering of Lake Michigan's water level. With lakeside vistas from their porches potentially spoiled by Greenhouse-world entrepreneurs, our neighbors have awakened to a basic fact of life—if you don't own the land, you can't influence how it is developed, except through a change of zoning laws.

What should prospective buyers of lakefront property do? We offer this concrete suggestion for self-interested action. On inland sites away from the shorelines of the Great Lakes, choose a lake that is fed by springs and inflowing streams to ensure better stability of water level. Locate a sector of lakeshore with deeper water just offshore. When the lake level drops, your dock will still provide a deepwater access channel for your motorboat, obviating the need for dragging it across an exposed mudflat. Another way of looking at this situation is that the extended overland portage carrying your kayak or canoe will add to your "wilderness experience." Select a lake basin with a relatively large area of water greater than twenty-five feet deep. The water volume of a larger lake will remain cooler and will hold more dissolved oxygen, a hedge against summer fish kills.

Gone fishing

"Structure." Our fishing buddies tell us that if we really want to catch the lunkers, the truly Big Fish, structure is everything. Structure means you have to think like a fish . . . if you want to find it on the other end of your fishing line! Fish are cold-blooded. Within a few tenths of a degree, their body temperatures are locked in at the temperature of the water in which they swim. This **thermal habitat** dictates how much and how fast a fish can eat, and this in turn determines how fast the fish will grow.[9]

Fish are picky about where they choose to live. This is not news—we've fished in lots of places where the fish weren't. Good fishermen learn enough about fish behavior to predict where they will catch their daily limit. They also target sites with suitable physical structure, for instance steep drop-offs along the lake bottom and submerged snags along shallow bay margins. Such aquatic sites are sheltered from large predators and have both food (in the form of plankton and smaller fish) and oxygen in plentiful supply.

Different groups of freshwater fish prefer different water temperatures, ranging from 50 to around 82°F. Fisheries biologist John Magnuson and his colleagues from the University of Wisconsin's Center for Limnology tell us that fish deliberately seek out the water with the temperature closest to their own optimum.[10] Given a suitable thermal habitat, fish live about 66 percent of the time in waters within 4°F of their preferred temperature and 90 percent of their time in waters within 7°F of their optimum. To find lake trout, cast your lures in very cold water measuring about 50°F. In the Great Lakes, lake whitefish and coho salmon school in cold waters of about 54 and 60°F, respectively. Cool-water species of walleye and yellow perch can be netted from waters of 68 and 73°F, respectively. The warm-water species of largemouth bass prefers its ideal of about 81°F.

Within the Great Lakes, both commercial and sport-fish species move with the seasons. Certain fish move vertically in the water column; others move laterally between deep open water and shallower coastal areas. They are thus adapted to the seasonal cycle of changing thermal structure. For example, cold-water fish like lake trout move vertically. In midwinter, lake trout swim beneath the frozen ice cover but stay in the upper lake water. In the summer, after the surface water warms up, they move to deeper, colder water. Warm-water fish move horizontally, migrating between shallow coastal bays and inland streams in winter and open water in the summer.

In a Greenhouse world, as the seasonal cycle of thermal structure changes, fish populations will have to respond to survive. Michael McCormick, a scientist at the Great Lakes Environmental Research Laboratory, National Oceanic and Atmospheric Administration (NOAA), has studied how Greenhouse-climate warming will affect Great Lakes thermal structures. He documented today's seasonal patterns in water temperature for southern Lake Michigan using buoys stationed along shipping lanes, which provided air temperatures as well as water temperatures from the lake surface down to about 160 feet depth. McCormick found that the most important structural dynamics occur in the surface water. Commonly, in winter the surface of Lake Michigan freezes over, capped with a continuous ice layer. As the air warms in springtime, the ice melts and surface water warms from 32 to 39°F, the temperature at which freshwater reaches its maximum density. This oxygen-rich, dense surface water sinks, causing the water column to turn over and to mix thoroughly. Continued surface warming causes the less dense, warmer water to form a cap over the colder, denser water below. Summer wind and waves churn the surface and deepen the thickness of this warm-water cap, sharpening the transition zone, or

thermocline, between the upper layer of warm, buoyant water and the lower layer of cold, dense water. As the air cools in autumn, surface water cools down to 39°F, producing a fall **overturn** in water circulation. Further cooling produces the winter pattern of temperature layering and lake surface covering of ice.

McCormick mimicked this seasonal pattern of lake dynamics using a computer model that simulated spring and fall overturns. Then he ran computer simulations for both GISS and GFDL scenarios of Greenhouse warming based on a doubling in atmospheric carbon dioxide. For southern Lake Michigan, these simulations projected summer air temperatures up to 16°F hotter than today, with winter temperatures increasing by 14°F. Temperatures in the uppermost 500 feet of Lake Michigan will also be warmer than they are today, by 3°F in late summer and 6°F in late winter. In the Greenhouse world, spring overturn will generally occur in early May, one and a half months earlier than today. Autumn overturn should be delayed at least one month, until early to mid-December. Thermocline layering will stretch from today's five months, mid-June through early November, to about seven and a half months in the future. The coldest waters, in February, will be 41°F. Lake Michigan will no longer consistently freeze over in winter, and the water layers may not regularly turn over in spring or fall either. Permanent thermoclines may develop in the deep lake basins, trapping stagnant bottom water. Lacking in dissolved oxygen, it will not be suitable for sustaining fish populations, decreasing the cold-water habitat.[11]

John Magnuson tells us the rest of the story. During most of each year, Lake Michigan is now much too cold for even the coldest-water lake trout to thrive.[12] For them, Magnuson predicts that long-term warming of the Great Lakes will trigger a cascading series of *positive* changes. The single-cell algae and other tiny plant life (phytoplankton) that form the base of the aquatic food chain could triple in biomass, doubling the harvest from fisheries located in shallow bays along the coastlines of Lakes Michigan, Erie, and Superior.[13] These fantastic estimates may be too good to be true, however, at least for the southern third of Lake Michigan and for Lake Erie, where massive algal blooms may deplete dissolved oxygen and result in fish kills. If water circulation in Lake Michigan is inhibited by the surface-capping effect of a permanent thermocline, the expansion of fishery stocks may also be limited by lack of dissolved oxygen in deep basin habitats.[14]

On the whole, nearly all native species of fish should respond aggressively to a longer growing season and the broadening range of water with suitable temperatures projected for a Greenhouse world. John Magnuson, Donald Meisner, and David Hill predict that Greenhouse-world environmental changes will warm Lake Michigan's too-cold waters enough to shift seasonal temperatures *into* the ideal ranges for cold-water populations of lake trout and coho salmon.[15] To make their estimates, Magnuson, Meisner, and Hill mapped the depth in the water column having the preferred temperature of each of four species of sport fish (figure 5-1). They also plotted the ranges in depths where those fish should spend between two-thirds (black pattern) and 90 percent (shaded pattern) of their time. They found that the combination of a bigger range of acceptable water column and longer seasons translates into a much more favorable thermal structure. The Greenhouse-world projections indicate a habitat expansion for lake trout by 2.7 times, for coho salmon by 1.4 times, and for yellow perch by 2.5 to 3.1 times, which could cause a yellow-perch population explosion. Summer temperatures will warm surface waters enough for even the most restricted fish, the largemouth bass.

With a copy of figure 5-1 taped to your boat's fish finder, you can use your LCD displays of water temperature and water depth to target the fish you want to catch. Or, if you prefer a low-tech approach to fishing, you can use these charts to figure out where the fish are most likely to be and maximize your chances of bringing home dinner the old-fashioned way! In a Greenhouse world, one of your favorite lifestyles—gone fishing—may be centered on one of the Great Lakes.

Winter wonderland?

To us, Michigan means a winter wonderland. Especially for Hazel, who grew up in Muskegon on the eastern shore of Lake Michigan, childhood remembrances of the 1960s and 1970s center on visions of snow forts, backyard ice rinks, sledding down sand dunes covered in deep snow, and parents using long-handled rakes to remove the heavy snow load off roofs. Boomer friends who remained in western Michigan say that winter weather isn't as it used to be when they were small. Today, cross-country skis stay tucked away inside closets most of the winter. The snow no longer falls thick enough or lasts long enough for good skiing or tobogganing.

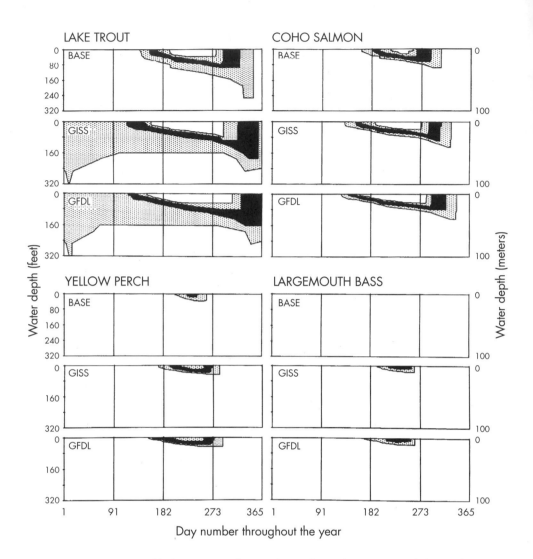

Figure 5-1. **Seasonal habitats in southern Lake Michigan for important fish populations, today and with Greenhouse-climate warming.** Fish seek out water at their preferred body temperature. If food availability and levels of dissolved oxygen remain adequate, they will spend 66 percent of their time within waters defined by the black band and 90 percent of their time within waters defined by the broader shaded zone. Today's habitats were simulated with the BASE computer model; Greenhouse-climate habitats with the GISS and GFDL models. (Adapted from John J. Magnuson, J. Donald Meisner, and David K. Hill, "Potential Changes in the Thermal Habitat of Great Lakes Fish after Global Climate Warming," *Transactions of the American Fisheries Society* 119 (1990): 254–264. Modified with permission of the American Fisheries Society.)

Residents of the Lower Peninsula have turned into wanna-be yoopers, longing to abandon their snow-bare winter landscapes for the groomed snowmobile trails of Michigan's Upper Peninsula.

We are now considering whether to return to the Great Lakes region for our own retirement, for part or all of the year. Will we find what we remember, the deep snow that draws us back to our roots? How will Greenhouse warming influence our prospects—and other Boomers' prospects—for future wintry wonderlands?

Hazel's hometown, Muskegon, boasts a more equable climate with milder temperatures year-round and with more precipitation than experienced at places farther inland. A band of this **lake-effect climate** usually extends no more than sixty miles inland, paralleling the eastern and southern shoreline of Lake Michigan (figure 5-2).[16] Widely known as a fruit belt, this elongate coastal strip supports extensive orchards of apples, peaches, and cherries, as well as vineyards for local wineries. It is also known for its snowstorms.

Figure 5-2. **The lake-effect snowbelt.** The stippled pattern, bordering eastern and southern shorelines around the Great Lakes, shows where heavy lake-effect snows fall today—an effect that global warming will probably strengthen. (Adapted from D. C. Norton and S. J. Bolsenga, "Spatiotemporal Trends in Lake Effect and Continental Snowfall in the Laurentian Great Lakes, 1951–1980," *Journal of Climate* 6 [1993]: 1943–1956.)

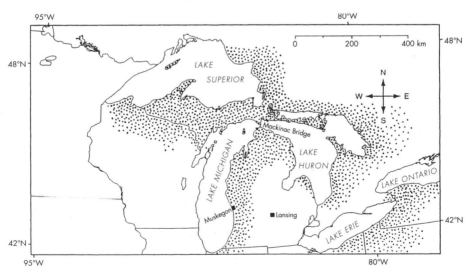

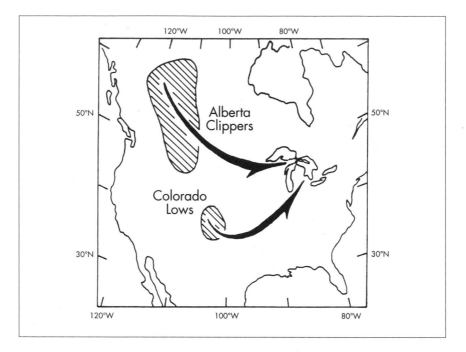

Figure 5-3. **Routes taken by major midwinter storm systems.** Colorado Lows and Alberta Clippers both bring snowfall to the Great Lakes region. With global warming, the Colorado Lows may bring rain or ice storms instead. (Adapted from S. N. Rodionov, "Association between Winter Precipitation and Water Level Fluctuations in the Great Lakes and Atmospheric Circulation Patterns," *Journal of Climate* 7 [1994]: 1693–1706.)

Climatologist Van Eichenlaub tells us more about such winter storms, which generate lake-effect snowfalls downwind of each of the Great Lakes, and he offers rules of thumb for forecasting them. Winter storms can occur any time from October through March. They require unusually cold periods with bitter, subfreezing temperatures and strong westerly winds, which have to travel at least fifty miles over the ice-free surface of a Great Lake. This fetch is the minimum distance for airflow to pick up enough moisture to accumulate as significant snowfall downwind in the coastal snowbelt. With a greater temperature difference between frigid air and warmer surface waters, there exists a greater chance for lake-effect snow. As Van Eichenlaub says, "If all these conditions are met, it is a good bet that you will need a snow shovel." Early to middle winter is the big season for this type of snowfall. Later, by February and March, lake surfaces have chilled down to freezing, and the lakes typically become covered by a sheet of ice. As the late-winter cap of ice cuts off the local moisture supply, the frequency and intensity of lake-effect snowfalls taper off.[17]

The big weather picture of figure 5-3 shows the two major storm tracks taken by low-pressure systems bringing midwinter snowfalls to the Great Lakes region.[18] Along the southern route, low-pressure systems called **Colorado Lows** form over the Four Corners area of the American Southwest. Shifting eastward, these growing storms pull in water vapor and heat energy from the Gulf of Mexico. Then, sweeping northward into cooler realms, Colorado Lows transporting this warm, moist air produce heavy snowfalls throughout the lower Midwest.

In contrast, northern low-pressure storm systems form from very cold, very dry air in the plains region east of the Canadian Rockies. The midwinter path of the high-altitude polar jet stream steers such fast-moving storms southeastward. As these moisture-starved **Alberta Clippers** track across the typically ice-covered waters of the Great Lakes, they bring little more than high winds and frigid temperatures, sometimes as cold as minus 40°F. However, should one of these wintertime Alberta Clippers pass over relatively warm, unfrozen waters of Lakes Superior and Michigan, the storm will sponge up heat and evaporating water vapor. The result can be a spectacular episode of lake-effect snowfall.

The decades of the 1960s and 1970s were indeed a time of heavy lake-effect snowfall in western Lower Michigan (figure 5-4). Those decades, however, were also anomalous. During the first part of the twentieth century, winters in Muskegon were not characterized by as heavy snowfall events, nor were the winters after the 1970s. In part, this spike during our Boomer childhood years may have been due to increases during this century in average atmospheric temperatures. But what about future winters in a Greenhouse world?

Certainly by the Boomer Breakpoint year 2025 people should experience firsthand the Greenhouse-world scenarios that call for much warmer and much wetter Midwestern winters. Beware, though, the statistics projecting a gradual warming trend "on average." We expect that the projected Greenhouse warming will indeed come about, but with *greater variability* between the kinds of frequent and extreme weather events. Warmer *averages* can result both from an overall shortening of the winter season and

THE LAKESHORE

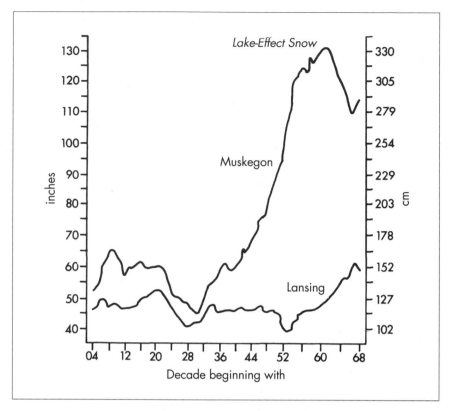

Figure 5-4. **Snowfall in Muskegon, Michigan, and Lansing, Michigan.**
Muskegon, on Lake Michigan's eastern shore, received an avalanche of lake-effect snow during the 1950s and 1960s compared to Lansing, 100 miles inland. (Adapted from Van L. Eichenlaub, *Weather and Climate of the Great Lakes Region* [South Bend, Ind.: University of Notre Dame Press, 1979].)

from a shifting balance between northern and southern sources for winter storms (figure 5-3), including an increased prominence of Colorado Lows.

In a Greenhouse world, residents of the Midwest should expect ever changing sequences of cold, penetrating rainfall, of paralyzing glaze ice, and of heavy, slushy snowfall.[19] Winter air temperatures will fluctuate in a broad range centered around 32°F, water's freezing temperature. The arrival of more January thaws and soaking winter rains may tax even the awesome capabilities of the snowmaking machines that maintain the artificial snow base on downhill ski runs. It's "snirt" at best (that's a gritty mix of snow and windblown dirt—a wintertime phenomenon known well to Minnesotans).

But the good snow, when it comes, should be glorious—if you live in the right place. On all the Great Lakes except perhaps northernmost Lake Superior, surface waters will not freeze over completely. As in the 1960s and 1970s (peak years of lake-effect snows), the Greenhouse-world combination of relatively warm, unfrozen water and frequent passage of Alberta Clippers should once again produce a winter wonderland. Exceptionally deep accumulations across the lake-effect snowbelt (figure 5-2) will invigorate the winter tourist season for Michigan's northwestern Lower Peninsula, the whole Upper Peninsula, and the Ontario sector lying to the east of Lake Superior and north of Lake Huron. If you choose to retire in one of these destinations for your winter lifestyle, consider investing in good snow tires, a vehicle with four-wheel drive, and a heavy-duty snowblower! For a low-tech solution, reconsider your traditional alternatives: cross-country skis or snowshoes, and a trusty snow shovel!

Forests in Peril?

We and many of our professional colleagues are forest historians by trade. We have in our bag of tricks many kinds of tools that enable us to read landscapes of both the past and the present and to learn how environments develop—how climates change, lake levels fluctuate, and soils form.[20] For example, using measuring tapes and laptop computers, we can sample the species composition and canopy structure for the kinds of trees that form today's forests. In order to track the history of these forests, we can sift through dusty archives, including naturalists' journals[21] and surveyors' diaries[22] dating back as much as two centuries. Pushing deeper into time, we can take tree cores, which record centuries of climate fluctuations in their yearly growth of tree rings. Mostly we take the long view, wearing the hats of **paleoecologists**. With coring devices, we bore through the mucky peats of wetland bogs and collect the ancient oozes preserved in lake-bottom muds. These lake and bog sediments contain microscopic clues about past landscapes, preserved in the form of pollen grains, fruits, and seeds shed by trees that, during past millennia, grew on the surrounding land. Our paleoecological time machine can take us back tens of thousands of years to decipher changes in vegetation and climate since the last ice age.[23]

Forest plots, tree rings, and plant fossils all yield glimpses of the dynamic processes involved in the life and death of forests. From the perspective of "tree time," we gain a better view of dynamic change in forest communities. Such *back*casting through forest history helps us as we change hats to those of **eco-futurists**. Now we must attempt to *fore*cast—to assess whether or not natural forests are in peril.

How do plants respond to climate change?

Each plant species has its own set of tolerances for coping with a stressful environment.[24] Some woody plants can tolerate extreme cold. Canadian conifers "cold-harden" by producing antifreeze within their cells that allows them to survive down to minus 85°F. Oaks, ashes, maples, and beech can resist freezing down to about minus 44°F. In contrast, palm trees are highly frost sensitive and grow only in subtropical to tropical climates. Maps portraying this fundamental relationship between climate and plant ranges are published online for all the important species of trees and shrubs native to North America.[25] These new biogeographic atlases document environmental requirements for 407 kinds of woody plants that together form the forests and shrubby vegetation stretching from the tropics to the northern treeline limit, adjacent to the arctic tundra grasslands. In the past, such knowledge of the tolerances of many kinds of plants to winter cold has led horticulturists to develop very practical maps. These maps depict cold-hardiness zones to guide landscapers and gardeners in selecting the varieties of ornamental plants and agricultural crops suitable for their region.[26]

As any gardener knows, most plants are also highly vulnerable to prolonged droughts. They may be more vulnerable during one stage of their life cycle, for example, germination of seeds. Plants may be less tolerant of extremes in either temperature or precipitation if they are near the margins of their natural geographic ranges or they are further stressed by atmospheric pollution or by insect or fungal pests.

If climate changes, either in long-term variability or in "average" conditions, key thresholds of living tolerances may be exceeded. The result may be a wholesale die-off or more gradual elimination of a species. Climate change can trigger unexpected responses in natural communities because of subtle effects—in the frequency, say, of recurring disturbances such as floods, windstorms, and wildfires. Particular disturbance events, including man-made changes such as timber harvests, offer "windows of opportunity" for the invasion of weedy plants into previously intact native vegetation.

In the Greenhouse world, three options exist for forests of the Great Lakes region.[27] These options also apply to the rest of North America. The first of these biological options is to *adapt*. In the "short" term, this means gradually changing the mix of species in a dynamic community, toward fewer conifers, say, and more hardwoods, over many generations. The fossil record shows us that, during past times of global warming, forest communities have taken many thousands of years to adjust.[28] Now the time frame has changed. Because of the accelerated pace of global warming, forests will have only the brief time span of the next several centuries. Will it be enough? The two remaining options are to *move it* or *lose it*.

How can trees move?

Trees don't just pack up their roots and travel on. Their migrations take many reproductive generations to complete because trees depend on their seeds for mobility. Whether dispersed by birds, small mammals, wind gusts, or flowing water, the seeds of forest trees must germinate and grow into the next generation of trees. In turn, these new trees become the advance guard that provides the next seed source to spread the species a little farther as it migrates across the landscape. Trees characteristic of the Great Lakes region today arrived there between 10,000 and 7,000 years ago. They moved northward from ice-age havens near the Gulf Coast, spreading at the rate of seven to eleven miles per century. Along its advancing margin, each vanguard population of trees produced the seeds for the next generation to leapfrog onward.[29]

Will these trees stay in the region much longer, or will they have to keep moving on to survive? Forest ecologist Margaret Davis of the University of Minnesota and her student, Catherine Zabinski, used the GISS and GFDL climate scenarios to project the geographic areas of suitable Greenhouse-world climate for four important Great Lakes tree species: eastern hemlock, yellow birch, sugar maple, and beech. Davis and Zabinski used computer simulation results for lowest January temperatures to project the future northern range boundaries of each forest species, based on their cold-hardiness tolerances. They used highest July temperatures to map the southernmost Greenhouse-world distributions, and they used annual precipitation estimates to determine the probable western limits, according to each species's drought tolerance.

The sad news is depicted in figure 5-5: we can expect *regional elimination of major tree species* within and to the south of the Great Lakes. To survive, these trees must move north some 300 to 550 miles during the next century, establishing themselves in new locations far north in eastern Canada. Although required for the Greenhouse world, this pace of migration *far* exceeds historical precedents, with the potential result that entire forests will die in place. Only a concerted planting program undertaken by the commercial wood-products industry and coordinated between U.S. and Canadian government forest services may yet rescue these important forest species.[30]

One optimistic note is that small forest preserves may persist along the shores of the Great Lakes because the local ameliorating lake effect will buffer extreme cold in winter, damp the extremes of summer heat, and contribute precipitation to keep soils from becoming too dry. As a result, shorelines of the Great Lakes will continue to be scenic, with their old-growth and biologically diverse forests of hemlock, yellow birch, beech, and sugar maple. In contrast, landlocked areas will become desolate scrublands as interior forests experience massive die-back. Forested vistas and water frontage along Great Lakes shorelines will become even more valuable in the near future, far exceeding $1,000 to $10,000 per foot of beach frontage, the price currently paid for coastal real estate in northern Lower Michigan.

Boom and Bust in Great Lakes Forestry

Dan Botkin, a professor of ecology at San Diego State University, did the impossible. He bottled up a virtual forest inside a computer and, in so doing, championed a fundamentally new way to look at nature.[31] For Botkin, the ecological process is everything. In 1970, he developed a computer model to link forest dynamics to environmental conditions. Called JABOWA-II after the initials of the three men who initially developed it, this

Figure 5-5. **Projected changes in distributions of hemlock, beech, sugar maple, and yellow birch under Greenhouse-world climate conditions.** Horizontal lines show present geographic ranges; vertical lines show potentially suitable ranges in GISS and GDFL scenarios. Bottom line: These vast northern forests may not survive *this century*. (Adapted from Margaret B. Davis and Catherine Zabinski, "Changes in geographic range resulting from Greenhouse warming: Effects on Biodiversity in Forests," pages 297-308, in R. L. Peters and T. E. Lovejoy, eds., *Global Warming and Biological Diversity* [New Haven: Yale University Press, 1992].)

GISS model GFDL model

HEMLOCK

BEECH

SUGAR MAPLE

YELLOW BIRCH

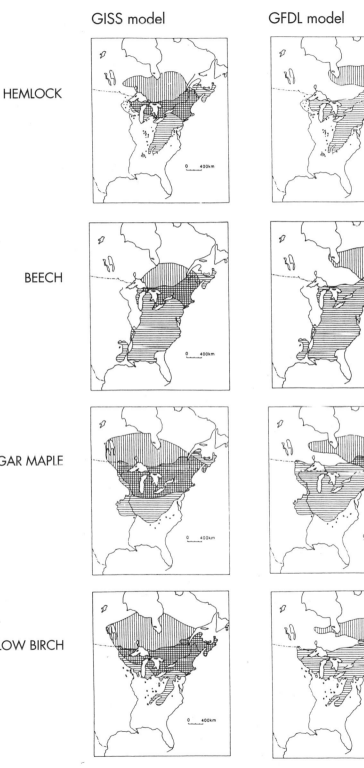

computer model uses mathematical equations to mimic the biological processes of life and death for thirty-four kinds of trees that form the Great Lakes forests. Simulated by the computer model, virtual seeds germinate, seedlings grow up into trees, compete with each other for sunlight and soil nutrients, and eventually die.[32]

To find out what we can expect to happen to Great Lakes forests under a business-as-usual scenario of Greenhouse warming, Botkin wedded his model of forest dynamics to the panorama of environmental changes projected with the GISS climate model.[33] He found that dramatic and enduring changes will happen to these regional forests corresponding with significant Boomer Breakpoints. Between the years 2010 and 2040, today's patchwork of forests and agricultural lands will shift toward a landscape mosaic dominated by cutover brushland and treeless bogs. These previously unexpected transformations will be accelerated by natural disturbances such as wildfire and by commercial activities including clear-cutting of the forest.

For example, today stands of scrubby jack pine trees grow on dry, coarse, sandy soils that are poor in nutrients. Stands of jack pine in northern Lower Michigan provide shelter and nesting sites for small populations of an endangered bird, the Kirtland's warbler. Botkin's model predicts that both the jack pines and the Kirtland's warbler are doomed to extinction by the first Boomer Breakpoint in the year 2010. Even though the pines are tolerant of fire, their sites will soon become too dry and summers too warm for their seedlings to germinate and grow after the next major wildfire or logging. Instead, white pine and red maple may temporarily replace the jack pine. Even these hardy species will be extirpated by the year 2040, and thereafter only a low growth of lichens, grasses, bracken fern, and blueberries will endure in treeless barrens on these extreme sandy sites. By the latter part of the twenty-first century, the Greenhouse-world climate will become hostile for all the northern pines. It will be too hot and dry not only for jack pine but also for red pine and white pine (the soon-to-be-endangered state tree of Michigan). Paradoxically, however, it will not yet be hot enough for southern pines to replace them by migrating north from the southeastern coastal plains. The great pineries of the Great Lakes region, which in the 1800s were the source of billions of board feet of lumber for building Midwestern cities, will be gone.[34]

For the Boundary Waters Canoe Area, Minnesota's million-acre wilderness located on the border between the United States and Canada, Botkin's

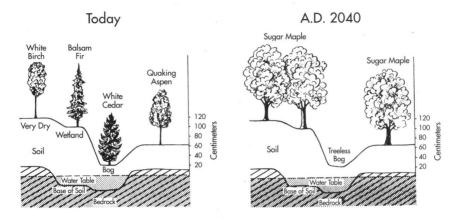

Figure 5-6. Forest canopy changes anticipated in the Great Lakes region. By the year 2040, forest landscapes will have changed dramatically as wetlands become barren of all trees, evergreens die off everywhere, and new hardwoods take over the higher ground. (Modified from fig. 21.4 in Daniel B. Botkin and R. A. Nisbet, "Projecting the Effects of Climate Change on Biological Diversity in Forests," pages 277–293, in R. L. Peters and T. E. Lovejoy, eds., *Global Warming and Biological Diversity* [New Haven: Yale University Press, 1992].)

forest simulation model predicts landscape-level changes.[35] In the uplands, dry knolls and moist hillslopes will become much drier. Some wetlands may dry out, allowing trees to establish in their moist depressions. Other bogs may persist, although reduced in acreage. Dropping water tables may dry up rivers that will no longer feed lakes. Because lakes, bogs, and floodplain swamps will shrink, they will no longer serve as natural firebreaks, and natural wildfires may no longer be held in check. Sugar maple will become the most important tree, replacing white birch on fertile soils, displacing balsam fir on midslopes, and outcompeting quaking aspen on thin, sandy soils (figure 5-6). Swamp trees of northern white cedar will slowly die out as wetlands dry up, and be replaced by species-poor blueberry thickets and sugar maple forest.

Commercial foresters may hasten the conversion of forest to brushland in the upper Midwest. Because tree vigor will diminish in the Greenhouse world, timber harvests will decline by as much as 50 percent. Since a forester's mind-set is geared to maximizing the "crop" of commercial wood harvested on either public or private land, the temptation will be to utilize what is available as soon as possible. According to Dan Botkin,

The greatest harvest to be obtained between today and year 2070 is the harvest one can obtain now. . . . Forests after today will only decline in total organic matter. Some may recover in the future, as species adapted to warmer and drier habitats migrate onto the land or are planted, but regrowth will be slow compared to human economic interests. This result is ironic. . . . [A] single-purpose forester would do exactly what the environmentalists would least like him to do. He would go out and cut all of his existing timber stands and try to sell the forest products before his competitors realized the same thing and flooded the market with timber.[36]

Thus, a spectacular pulse of commercial logging can be expected on private lands during the next ten years. If they are shortsighted and unprepared for Greenhouse-world changes, some forest products companies may choose to take what they think will be the biggest profits now, cashing in their stakes in a sucker boom just before the bust, causing a near-term over-supply far exceeding demand. Following this decade-long glut, however, will be a major slump in lumber supply. As demand accelerates above dwindling supply, construction costs can be expected to rise sharply, causing the prices of Boomer retirement homes to soar. After the Boomer Breakpoint year 2010, only ecologically savvy foresters who have delayed harvesting will be prepared to cash in on the real economic boom that will follow the bust. While selectively harvesting mature trees and maintaining the structural integrity of old-growth forests, these eco-entrepreneurs will be able to satisfy (with considerable profit) the peak demand for Boomer retirement housing between 2010 and 2030. Economically viable and ecologically sustainable forestry practices can maintain renewable supplies of timber, along with a landscape patchwork preserving migrational corridors for trees seeking northern Greenhouse-world refuges. Rather than being a strictly rear-guard action, such practices might even *enhance* the biological diversity of resident, native plant and animal species.[37]

A Hunter's Wild Card

As Boomers retire in droves between 2010 and 2030 (figure 2-2), it is very possible that extensive logging, combined with massive tree die-offs induced by Greenhouse warming, will convert vast landlocked areas of forest into more open, treeless plains. Forest zones should persist in coastal corridors with a lake-effect climate. Because of these landscape changes, the Great Lakes region should become a mecca for sportsmen and sportswomen. The forests' ecological replacements—weedy meadows, shrubby thickets, and boggy muskegs—may persist for centuries. How nice for the right sort of entrepreneur or nature lover. Some state agencies, such as the Michigan Department of Natural Resources (DNR), are already tapping into the proverbial silver lining found in every Greenhouse cloud, providing creative value-added services online to their citizen constituency.

Opportunities will abound here. For example, in a Greenhouse world increasingly swept by fire (either by accident or by prescribed burning), natural biological communities that regenerate afterward offer new recreational opportunities. The Michigan DNR maintains a list of the previous year's burn sites on state-owned land as a "Mushroom Hunting Guide."[38] These forest openings are good collecting sites not only for morel mushrooms but also for wild blueberries. Attracted to this new food resource, wildlife including partridge, wild turkey, and white-tailed deer will also flock to these new forest-edge and barren habitats. Future wildlife populations should grow to exceed even the bumper years of the 1930s, during which wildlife multiplied to occupy vast areas of suitable new habitat in the form of extensive cut-over brushlands left by the first wave of logging at the beginning of the twentieth century. In a Greenhouse world, enhanced lake-effect snows will help preserve coastal corridors of species-rich forests. But the deep snowpack may force white-tailed deer to overwinter in deeryards far inland from the Great Lakes shorelines.

Whether your goal is leisure at the lakeshore, great sportfishing, berry picking, mushroom collecting, or wild card hunting, the Great Lakes region will continue to be an attractive retirement destination during the twenty-first century. Of course, the lifestyle choices that will draw people to such an area go well beyond the prospects for good fishing and hunting and sitting in the sun. For some Boomers, the real attraction lies in a desire to be connected to nature and to live in the presence of vibrant biological diversity. At least along the lakeshore corridors, we will continue to reside near natural, old-growth forests. Like us, if we choose wisely, trees will have the right habitat in which to adjust and even flourish with Greenhouse-world warming.

CHAPTER

6

The Mountains

The stretch of Blue Ridge Mountains extending from Virginia to northern Georgia are renowned for their scenic beauty. Recently, we paid a visit to friends who live near Black Mountain, northeast of Asheville, North Carolina, near the Blue Ridge Parkway. Our friends' home sits on a seemingly pristine, rustic, wooded mountainside, in a new subdivision with each house centered on a several-acre parcel of land, overlooking spectacular views of surrounding hillsides and sheltered coves. After complimenting our friends on the fabulous setting, we were puzzled by their dissatisfaction. Their main complaint involved the lack of an enclosed garage. Inquiring further, we were told that it was not the inconvenience of slogging between house and car in the rain or snow that upset them. What bothered them most was the myriad of tiny rust spots that had appeared suddenly, pitting their car's finish after only a few months of living on top of the world. The rust spots, it turned out, had been etched by acid rain.

Wellness in the Back of Beyond

How will ever increasing levels of air pollution impact our quality of life in the Greenhouse world? We are first enticed as tourists, then drawn back as retiring Boomers, to the unique lifestyles offered by mountain destinations. As the southern highlander Horace Kephart once said, if we seek solitude, we can find it in the majestic panoramas of blue ridges that lie "back of beyond."[1] Boomers seek a sense of wellness, of rediscovering a personal center, as we stroll along quiet walkways shaded by overhanging hemlock boughs. From a lofty perch, we can ponder solutions for the environmental conundrums ahead.

How will Greenhouse warming alter what we value most? What will happen to the enchanting, misty vistas and neverending skylines unfolding before us as we walk the Appalachian Trail? Will brook trout still rise to the Thunderhead dry fly cast on a gossamer line over a whitewater cascade? Tucked away in a Smoky Mountain holler, which kinds of plants might wink out, lost forever to our wildflower pilgrimages? Which colors will fade for "leaf peepers" accustomed to Jack Frost's blazing palette of autumn foliage? Can ecological consciousness represent the next step for improving our world?

In this chapter, we address issues of environmental quality as a prerequisite for achieving a fuller quality of life. We explore the threats of acid rain and ozone pollution, changing weather patterns, extinctions, and retreating ecotones that endanger mountain vistas, clean air, pure water, and the biological treasures we value most within the mountain wilderness.

What is acid rain?

Acid rain is today a major problem in the southern Appalachian Mountains. A byproduct of atmospheric pollution, **acid rain** includes both rainfall from clouds and fog drip, the condensation of water vapor from wispy clouds that cloak mountain summits.[2] In the Great Smoky Mountains, park ranger Steve Moore has said, "Some rain goes down to a pH level of 3.0, which is [the same acid level as] vinegar. We have recorded cloud pH of 2.4. That pH level will dissolve a hot dog . . . .When acid rain gets into the streams, it lowers the pH and has, in some streams, killed all aquatic life."[3] The average acidity of rainfall in the Great Smoky Mountains National Park is pH 4.5, up to ten times more acidic than normal rainfall. Water in clouds of the Smokies averages pH 3.5, which is ten times again more acidic.

According to the Southern Appalachian Assessment (SAA), **sulfur dioxide** is primarily responsible for making rain acidic. Sulfur dioxide, a gas produced from traces of sulfur in fossil fuels, is changed into particulate form in the atmosphere. These highly acidic sulfate particles are scavenged from the air by falling raindrops. Acid rain then falls everywhere, etching paint from the exteriors of houses and cars. Deposited on leaves of green plants, sulfate particles are taken into plant tissues and cause irreversible damage. When rained out onto the landscape, sulfates add to the chemicals dissolved in and carried along by streams. Over time they make the water turn acid, particularly in the headwaters of streams in the Great Smoky Mountains and in lakes in the Adirondack Park of upstate New York.

A second major contributor to acid rain, as well as to ground-level ozone, is the gas **nitrogen oxide**. Produced as a byproduct of electrical power generation and by internal combustion engines in vehicles, nitrogen oxide and its derivatives are particularly potent pollutants affecting air quality. The resulting chemical nitrates, deposited on the ground surface, may reside in soils for relatively long periods of time and then redissolve into stream water. High-elevation soils in the Smokies have reached advanced stages of nitrogen saturation, causing toxic aluminum to leach into streams and lakes and be absorbed by terrestrial and aquatic life. Acid rain can thus affect plant and animal life long after the initial rain-out of atmospheric pollutants.[4]

Ozone

Ground-level **ozone**, a molecule with three atoms of oxygen, is created as gaseous nitrogen oxides mix with volatile organic compounds (natural or man-made) in sunlight. Researcher Jim Renfro observed that high gaseous concentrations of ozone pollution can be trapped within the mountains, particularly during episodes of air stagnation associated with summertime high-pressure weather systems.[5] Visitors to the park are frequently exposed to these high ozone levels. During the summer of 1998, forty-three days of unhealthy ozone levels, exceeding federal clean-air standards, were monitored in the Great Smoky Mountains National Park. Ozone at such concentrations can be a powerful respiratory irritant causing coughing and chest pains, and potentially damaging lung tissue.

Ozone can be life-threatening not only for us but also for other species. Certain trees, including white pine, tulip tree, and sassafras, are especially sensitive to ground-level ozone. Ozone enters pine needles through pinprick-size openings called stomata and kills internal leaf tissues, leaving the pine needles blackened and covered with stippled spots. This type of biochemical injury has been found throughout the southern Appalachian

Mountains. Within the Great Smoky Mountains National Park, Jim Renfro reported that ozone damage to trees and wildflowers appears to be most severe at high elevations. He documented that 21 species of trees, 15 types of herbs, 9 kinds of shrubs, 3 species of vines, 1 species of fern, and one type of grass showed signs of leaf damage from exposure to elevated ozone levels. Already, 90 plant species exhibit foliage symptoms of ozone "burn."[6] This biological evidence indicates that atmospheric ozone pollution is becoming widespread even within wilderness areas, and that numerous plant species across many natural communities are being affected.

The Southern Appalachian Mountain Initiative (SAMI), a partnership of eight southeastern states and three federal agencies, is developing regional strategies to protect southern Appalachian natural resources from deteriorating air quality. Its voluntary mission should result in a final report of recommendations to be published in 2001. Policy makers may or may not then use the report to enact pollution-control strategies that could reduce future acid rain deposition, ozone exposures, and regional haze.[7] In the meantime, except during extreme drought years (when plants take in less ground-level ozone), the scenic beauty of the Great Smoky Mountains and of other national parks and forests throughout the eastern United States will continue to be diminished by the tarnish of ozone damage.

Endangered vistas

Aerosols composed of sulfates, nitrates, volatile organic compounds, coarse dust, and soot scatter sunlight and thus create a haze that obscures distant objects from view.[8] Since 1950, in the central and southern Appalachian Mountains, they have significantly degraded air quality and decreased overall visibility by 60 percent (figure 6-1).[9] Today, visibility in the Great Smoky Mountains has decreased from a typical distance of ninety-three miles to a mere twenty-two miles. On hazy days with high sulfur pollution and air stagnation, views can be restricted to a distance of only one mile. Unfortunately, visibility is poorest during the humid spring and summer months, when visitation along the Blue Ridge Parkway and in the Great Smoky Mountains National Park is at its highest (figure 6-1). You can check out the hazy scene for yourself by viewing the highest ridgeline in the southern Appalachian Mountains, from Thunderhead Mountain to Mount LeConte, via the online

Figure 6-1. **Good and bad visibility days, Great Smoky Mountains National Park.** Views from the Look Rock Tower, looking across the Smokies of Tennessee and North Carolina, show the extent to which sulfur dioxide pollution in the atmosphere can reduce visibility, from 100 miles on a good day down to 20 miles on a typical spring or summer day. (From U.S. National Park Service, Nature Net, 1998.)

Smokies camera set up on the Look Rock Tower. For comparable cam shots showing ozone pollution's impact at other National Parks, check out the Nature Net for localities in California, Maine, Texas, and Utah.[10]

Most sulfur dioxide gas and sulfate particles are emitted in summer, from power-plant combustion of fossil fuels to produce electrical power for air conditioning and from smog produced by automobile exhaust. Airborne sulfur dioxide that reaches the southern Appalachian Mountains comes both from within the southeastern United States and from other sources beyond, in the Midwest and Northeast. Sulfate particles scatter light and are responsible for up to 75 percent of the summertime loss in visibility.[11]

Nationally, emissions of sulfur dioxide in both gaseous and particulate form increased from the 1940s to the 1970s. Then, because of legislation that reduced allowable emissions, atmospheric sulfur dioxide generally returned to 1940 levels. In the southern Appalachian region, however, sulfur concentrations actually increased 25 percent between 1984 and 1996, as a result of growing human populations and increasing energy demands.[12] Emissions of nitrogen oxides have also increased steadily since 1940. In the future, production of these airborne pollutants is expected to increase as vehicle use increases and as demand for electricity grows along with the human population and warming climate.

Given a 50 percent reduction of sulfur dioxide emissions, if we collectively behave according to the provisions of the 1990 Clean Air Act, visibility in the southern Appalachian Mountains should improve somewhat by the Boomer Breakpoint year 2010. The estimated improvement of only three to seven miles, however, may not make much difference in the haze-shrouded mountain views afforded to the average visitor in summer months.[13]

In the spring of 1999, anomalously high values of sulfur dioxide and other atmospheric pollutants, as yet unexplained, were reported in the Great Smoky Mountains National Park. On Earth Day, April 22, 1999, Vice President Al Gore called for additional reductions in the emission of atmospheric pollutants in order to restore air quality to 1940 levels by the year 2064, which will be *almost too late for Boomers* if indeed it happens.

Fishing holes gone sour

Endangering our views of mountain landscapes, of course, is not the only way that air pollution is affecting the quality of living and retirement possibilities in the Appalachian Mountains. Other effects, even more insidious, include acidification of streams, especially in the headwaters.

According to the 1995 *Acid Deposition Standard Feasibility Study* by the Environmental Protection Agency (EPA), the regions of the United States most at risk from continued acid deposition are located in the eastern part of the country.[14] Lakes and streams in most of the Appalachian Mountain chain, from the Adirondacks in New York to the southern Blue Ridge in Georgia, will be hit hard. Since 1976, streams in the Great Smoky Mountains National Park have already increased as much as three times in their average water acidity.[15] It is currently estimated that in the mid-Appalachian region, about 30 percent of stream reaches studied are likely to become acidic during the worst rainfall episodes, seven times more than the number of stream reaches that are currently chronically acidic.

Fish that may die when streams become too acidic include not only the native brook trout but also the introduced rainbow and brown trout, all of which are highly prized by fly fishermen and are a major drawing card to national parks in the eastern mountains. Dace, minnows, and the food base of aquatic insects begin to disappear with rising acidification.

Brookies

Known locally in the southern Appalachians as "spec" or "brookies," brook trout are easily identified by their distinctive light-colored spots, which are speckled over a dark body background, combined with worm-shaped color markings along their backs.[16] Technicolor populations of brook trout require the clear, unpolluted water of cold, fast-flowing streams.[17] During the growing season, brookies find their optimal temperature of 55 to 65°F in the uppermost headwaters of Smoky Mountain streams. Protected as they lie buried by stream gravel, incubating trout eggs are not damaged by near-freezing winter temperatures.

Brookies favor slightly acidic water. Should the pH drop much below 6.0, however, fish eggs and juveniles may die. Even brief exposures to heavy pulses of acid precipitation, leading to pH levels as low as 3.6 for as little as one week, will kill brook trout fingerlings. In winter, stream pH values sometimes reach levels in which brook trout fry cannot survive. Under such conditions, older trout become vulnerable to internal parasites and may fail to reproduce.[18]

At the beginning of the twentieth century, brook trout thrived in streams down to an elevation of 1,600 feet. However, logging of many watersheds in the early 1900s stripped away the cooling shade of creekside forests, resulting in warmer stream water. Mud eroding from denuded hill-slopes silted over the gravel channels necessary for spawning. With the deliberate introduction of rainbow trout around 1900 and brown trout during the 1950s, the native brook trout faced new competition for remaining stream habitat. Today, populations of native brookies survive only above 3,000 feet. This restriction in habitat by more than 75 percent means that they are faced with possible extinction because of acidifying habitats, warming water temperatures, and increased competition from introduced trout species. The National Park Service maintains an aggressive restocking effort as part of its Restoration Program for Native Brook Trout.[19] As fisheries biologist Steve Moore has said, "The Smoky Mountains are perhaps the best-known refuge for this genetically unique subspecies of brook trout. . . . Once these natural resources are gone, there's no way you can buy them back."[20]

It's an uphill battle to say the least. In the twenty-first century, the already endangered brook trout will be even more likely to go extinct. The combination of appreciably warmer summers and increasingly acidic stream water will threaten their survival on all but the highest summits of the Appalachian Mountains. Greenhouse-world projections call for fishing holes to continue going sour, with extirpation of brook trout from at least half of their already diminished range in the southern Appalachians.[21] As a bottom-line message for Boomers hoping to match their fishing skills against wily and wary brook trout, the uncertain fate of brookies brings home a tiny part of the Greenhouse-world reality. In fact many native species are now vulnerable to extinction—so many that biologists are still racing in the Smokies to discover and count as many as possible, before they are gone forever.

Paradox of the Tourists

We are doing it to ourselves! Twelve million tourists visited the Great Smoky Mountains National Park in 1999.[22] Tourists throng to the outlying towns of Pigeon Forge and Gatlinburg, Tennessee, a sprawling urban tangle of congested streets, country and western music theaters, and massive factory outlet malls. Casinos beckon travelers to trek across the mountains to the Qualla Indian Reservation in Cherokee, North Carolina. New six- and eight lane highway corridors funnel tourists from Interstates 40 and 75. As traffic congestion approaches gridlock on weekends, pollution levels from car exhaust inexorably increase.

Sandwiched between the tourist-trap extremes of Gatlinburg and Cherokee lies the Great Smoky Mountains International Biosphere Reserve.[23] The Smokies are one of our great national treasures. Imagine hiking up from the western park boundary. You climb to the summit of the highest peak you see. Before you lie more than 100,000 acres of old-growth forest, composed of trees that are not only very old but are also very big! The Smokies are home to twenty-one record trees, more species of national champions recognized than in any other American national park or forest.[24] From your perch on the first ridge, you hike on eastward to the tallest peak on the next ridgeline. On your adventure trek, you see *no* houses—not even one house festooned by the urban wildflower, the TV-satellite dish. Much of your trail winds through virgin wilderness, trees never cut by logger's ax. From your next vantage point, an aerie above 6,600 feet, locate your next trailhead and hike to the third, distant ridgeline. Your west-to-east journey covers more than fifty map miles through one of the world's truly great temperate rain forests. Wilderness found!

The tourists' paradox remains ironic: the more we love the southern Appalachian Mountains, the more we are loving them to death. Many attributes of park and wilderness areas that are valuable to the enjoyment of recreation seekers are also of vital importance to ecosystem health and biological sustainability. Clarity of the air through which we hope to view majestic blue-ridge vistas, well-stocked populations of trout in whitewater streams, and the soothing visual relief of attractive foliage from early spring leaf-out through the fall color show are all attributes of nature that we value for their aesthetic qualities. These attributes of the natural setting also indicate an ecologically high quality of life.

Diminishment of these natural attributes through degradation of their environment may call into question the appropriateness of continued overvisitation to national parks such as the Great Smokies as well as continued expansion of their gateway communities. Times of the year that are considered most desirable to visit the mountains are becoming the most vulnerable to pollution stress. Regulation of automobile access to the national park, not only to relieve traffic congestion but also to reduce air and noise pollution, is one possible way to alleviate some of the stress on natural ecosystems. Knowledge of Greenhouse-world climate trends can help us in making such choices, as we are better able to understand the interactions among climate change, pollution stress, and the ecologically desirable qualities of clean air, pure water, and biological diversity.

Wildflower pilgrims

Visitors to the Great Smoky Mountains National Park frequently stop at the Carlos Campbell Overlook just inside the park boundary on U.S. Route 441 above Gatlinburg. This vantage point offers a spectacular view across the Sugarlands Valley to the panorama of majestic Mount LeConte.[25] Today the vegetation is a mosaic of mixed hardwoods in sheltered valleys or "coves" at low elevations, oak forest on middle-elevation, south-facing slopes, and spruce-fir forest on mountain peaks above 4,500 feet elevation. Patches of pine forest and rhododendron "balds" occupy the most exposed ridges at middle elevations. Because of the relatively large elevational range, extending from about 1,000 feet to more than 6,000 feet above sea level, the varied aspect of the great number of folded ridges, and the great length of time that these 200-million-year-old mountains have been accumulating species, the Great Smoky Mountains are home to some of the most diverse natural biological communities in North America.[26]

Using tattered copies of our favorite books for identifying wildflowers, shrubs, and trees of the Great Smoky Mountains, each winter we plan out our weekend hiking schedule for the spring season. We hover over the spaghetti-like contour lines of a topographic map tracing out prospective trails. Just as Edwin Way Teal once followed spring from the Deep South to northern latitudes, we will see springtime in different stages as we hike from low to high elevations.[27] As connoisseurs of vernal blooms, we hope to see firsthand a succession of wildflowers, each species appearing in its own predictable time as spring unfolds in the mountains. In February we look for the brief show of bloodroot, hepatica, trout lily, and spring beauty forming a delicate carpet beneath the cove hardwoods. In April and May we await the brilliant color shows of trillium, wild geranium, and fire pink in hardwood forests at mid-elevations. In June and July, craggy knolls at high elevations are ablaze with the intense reds, oranges, and yellows of flame azaleas, intermixed with the equally showy masses of pink and purple mountain laurel and rhododendron.

Literally millions of other wildflower pilgrims also plan their annual return to the mountains. A fifty-year tradition in the Smokies, the *Annual Spring Wildflower Pilgrimage* is a three-day program held in the third week of April and consisting of nearly one hundred guided tours, including nature hikes, motorcades, and photography workshops.[28] The Southern Appalachian Man and the Biosphere (SAMAB) Web site offers a new Web page to help wildflower pilgrims plan their springtime touring agendas: *Watching Wildflowers* describes some of the finest natural gardens and wildflower demonstration areas available in the southern Appalachian Mountains.[29]

Timing your time-share resort

Wildflower pilgrimages in the Smokies are a major tourist attraction and provide a compelling reason for visiting the heartland of the eastern deciduous forest. Boomers are increasingly becoming stakeholders in our ecological future as we shell out hard cash for lifetime investments at a specific two-week interval in time-share condos and invest in vacation hideaways intended for retirement homes. But can we assume that what we see today is what will be there for us in the next twenty years? Perhaps, as environmental journalist Steve Nash suggests, we all need a forward-looking owner's manual for the Blue Ridge Mountains for the year 2020.[30]

In the Greenhouse world, the face of Mount LeConte in the Great Smoky Mountains will be dramatically transformed. By the Boomer Breakpoint in 2010, much of the present spruce-fir zone will be changed to northern hardwoods because of the dieback of Fraser fir. This native and globally rare species of fir tree has succumbed in recent years to attack by the balsam wooly adelgid, an insect that was inadvertently introduced on nursery stock in the 1950s. By the Boomer Breakpoint of 2025, the introduced European gypsy moth will make its way south from the northern Appalachians, killing back many of the old-growth oak trees.[31] By the last Boomer Breakpoint of 2070, climate warming will alter the plant hardiness zones, causing shifts in the ranges of many of the remaining native species of trees and their associated shrubs, herbs, mammals, reptiles, invertebrates, and fungi.[32]

The U.S. Environmental Protection Agency's global warming Web site offers the near-term environmental prognosis for your mountain escape, your particular state, or favorite national park.[33] The hard questions still remain, however. Just *which* two weeks of time-share, a time slot locked in every year for the rest of our lives, should we buy to maximize our spring wildflower experience? Will today's best flower shows still take place during the same two weeks when we retire? How should we time our time-share resort?

Let's start with the basics. What do we need to know? What controls the timing of the spring wildflower show? Is it temperature or day length? In the Greenhouse world will we see longer flower displays or only frost-nipped buds? How dependent are the spring wildflowers on the rest of the forest environment? What happens if the overstory trees die out in place and canopy-forming species migrate out *from over* the flowers? Will wildflower species migrate also to keep up with the pace of Greenhouse-climate change, or is their growth and dispersal too slow to escape to cooler sites located farther up the mountainside?

Wildflower life strategies

Spring wildflowers are specialists, and plant physiologists David Hicks and Brian Chabot understand their life strategies.[34] These two botanists tell us that wildflowers like to grow in the thick, moist soil duff that accumulates on the forest floor beneath hardwood trees. Vernal wildflowers are active during favorable periods of high sunlight, abundant soil water, and available nutrients that characteristically occur each springtime just before leaf-out in the overhead tree canopy. Spring ephemeral wildflowers capitalize on brief windows of opportunity for vegetative growth, flowering, and seed set. These plants may overwinter as small whorled rosettes of leaves, or they may remain dormant below ground as buried stems and root systems. As daytime temperatures rise above 40°F, each plant species is clued in to a slightly different temperature signal. Accumulating "heat units" mark the species-specific start of the new growing season. Spreading wildflower leaves intercept the direct sunlight penetrating through the stark leafless boughs of deciduous trees. The spring flush of flowers tracks the shifting angle of the sun, with flower displays expanding first across sunlit south-facing hill slopes, then progressing to more shaded settings around the mountainside. Each spring the wave of blooms creeps up the mountains as daytime temperatures warm; the temperature gradient of the mountains themselves imposes a regime of progressively cooler starting conditions at ever higher elevations.[35] The whole show takes about two weeks at any one location, but shifts with altitude over as much as ten weeks. The intrepid botanist can follow the spring for the same plant species in the Smokies, tracking it first in March within valleys near Sugarlands and finishing in May in high-elevation mountain gaps along the road to Clingman's Dome.

Firmly rooted in place, the populations of spring wildflowers are not as ephemeral as are their flower displays. Trilliums take over a decade between the time their seeds germinate until their first flowering. Their underground stems grow incrementally each year, and by counting the gnarled nodes of trillium and ginseng rhizomes, botanists have documented some individual plants that are well over fifty years old.

The beginning of spring is gradual for wildflowers. But its end is abrupt, coinciding with leaf-out among the hardwood trees, which then block sunlight from reaching the forest floor. The ephemeral wildflowers quickly go to seed and become dormant for yet another year, until their next Appalachian spring.[36]

Where will all the wildflowers go?

What is likely to happen to the spring wildflowers in the Greenhouse world? We don't really know, but three ecologists, Albert Meier, Susan Bratton, and D. C. Duffy, give us some clues based on their studies of wildflower diversity in the southern Appalachian Mountains. Meier, Bratton, and Duffy found that the number of wildflower species on a particular Appalachian mountainside varies with the frequency, intensity, and kind of disturbance they experience. Most kinds of spring ephemeral wildflowers grow beneath the canopies of old-growth forests, fewer species grow in forests that have been selectively logged, and the fewest are found in areas from which the forests have been clear-cut. Meier, Bratton, and Duffy project that, because of local wildflower extinctions linked to traumatic changes in the forest floor environment, at least a century may elapse before wildflower diversity can be restored to clear-cut lands. The habitat changes caused by logging include exposing the soil to erosion and to the desiccating summer heat. Once the soil litter is stripped down to bare mineral soil, mature plants and their seeds die and are quickly replaced by weedy herbs that can invade such disturbed sites. Even in the most favorable forested locales, many species of spring wildflowers grow slowly and with limited reproductive success. Their lack of effective seed dispersal also restricts their rate and potential range of geographic expansion. For example, the highly treasured medicinal wildflower ginseng drops its seeds within *only two feet* of the mother plant.[37]

In the Greenhouse world, the expected die-off of many trees may create effects similar to logging. The death or removal of tall canopy trees from the forest will create holes in the leafy canopy through which intense summer sunlight will penetrate and within which wildfires will be more likely to start. Such changes in the microclimate of the forest floor will be detrimental to cherished but highly vulnerable spring ephemerals including trillium, trout lily, doll's eyes, ginseng, may apples, dutchman's britches, and twisted stalk.

In the southern Appalachian Mountains, springtime warming will begin in January in the Greenhouse world, as much as *six to eight weeks earlier* than in the past. Leaf-out of the deciduous forest canopy will take place perhaps four weeks earlier. Instead of peak wildflower displays in April, early to middle March will be prime time for wildflower pilgrims in the Smokies. Eco-savvy Boomer botanists will invest now in "March futures" and thus will be able to lock in, at today's lower prices, those precious time-share weeks during what will become peak wildflower season in future decades.

Leaf peepers

What about equal time for tracking the autumn?[38] Fall leaf-peepers are at least as plentiful as spring wildflower seekers. What controls the timing and intensity of fall leaf colors: day length, moisture levels, or the onset of rapidly cooling temperatures? With range pullbacks for many deciduous trees, how will the panoply of autumn colors be affected in the Greenhouse world? Will we be able to enjoy the reds and oranges of sugar and red maples, the yellows of birches, aspens, and hickories, the browns of oaks, and even the verdant greens of hemlock and pines? Or will parts of the color spectrum begin to wink out as certain kinds of trees die off? Can we predict future areas of survival for variegated, mixed-species forests in order to invest in retirement destinations that will continue to have a good fall color show?

The autumn show of leaf colors usually peaks during the middle of October in the Great Smoky Mountains. Deciduous trees are sensitive to signals from two environmental cues: first, a decrease to less than twelve hours of sunlight each day reduces the level of photosynthesis; second, exposure to freezing temperatures shuts down the plant-food factories altogether.[39] The first environmental cue starts the gradual process of changing the blend of leaf colors. The second cue triggers the spectacular peak in colors, as the green plant pigment chlorophyll is eliminated and green no longer masks the colors of other leaf pigments. Leaf colors turn red and orange from residual carotene pigment, yellow from xanthophyll, and purple from anthocyanin. A hard freeze closes off sap supply from the twigs to the leaves, which form plugs called abscission layers at their leaf bases. The disconnected and obsolete leaves are then easily shaken off the trees by wind and rain.

In the twenty-first century, autumn foliage color will still begin about October 15 in the Smokies. Greenhouse warming won't change the celestial movement of Earth about the Sun that in turn causes day length to shorten.

In the Greenhouse world, however, the onset of freezing temperatures should be delayed until well into November in the southern Appalachians. The result will be a longer fall color show, with peak display taking place much later. Buying into a time-share condo for early November would thus be a long-term good value for frugal leaf peepers.

The *quality* of autumn foliage colors is as important to leaf peepers as the *timing* of peak color show. Will the intensity of color and the breadth of the color spectrum be affected by Greenhouse warming? The key to color intensity in autumn leaves is the rainfall and soil moisture levels of the preceding summer. In the Smokies, for example, the droughty summer of 1999 seriously depleted moisture in the soil, thus reducing upward sap flow through tree trunks and limiting how much starch was produced by plants through photosynthesis. By mid-October, deciduous trees began to turn a somber brown and to drop their leaves early. Pretty colors developed only on hardwood trees high on the mountainsides, where the trees had taken in extra moisture gleaned from passing clouds. A dramatic cold snap in early November triggered a brief but subdued fall color show during that warm and dry La Niña year.

Both GISS and GFDL climate models have the potential to indicate changes in the intensity of leaf color that we can anticipate in the future. The GISS model forecasts somewhat warmer and wetter summers, which would produce brilliant fall displays on par with those of most recent years. The more extreme GFDL model, however, projects much hotter summers with much depleted soil moisture. This scenario indicates that in the Greenhouse world, the trend will be toward lackluster leaf displays. For now, we can more confidently predict timing than intensity. If you end up in the right time-share slot but the colors fizzle, at least you'll be in the mountains.

The Dreaded E-Word, Extinction

Some naturalists view the landscape patchwork of varied hues in fall foliage as one conspicuous measure of rich biological diversity. A dazzling, colorful quiltwork of intermixed hues reflects the extraordinary number of tree species that share the forest canopy across the Great Smoky Mountains. What changes should we expect in landscape variety in the Greenhouse world? How will climate change alter the mix of species in the future—will we have to face the specter of the dreaded e-word, **extinction**?

Two forest ecologists, Louis Iverson and Anantha Prasad, both at the Northeastern Research Station of the U.S. Forest Service in Delaware, Ohio, have addressed the question of future extinctions for eastern North American trees. They used five computer projections, including the GISS and GFDL scenarios, to characterize potential shifts in geographic range and abundance for eighty tree species that will have to adjust to Greenhouse-world climate change.

Iverson and Prasad forecast that the southern Appalachians will lose many kinds of hardwood trees, including sugar maple, basswood, redbud, and yellow birch, as well as certain key evergreens such as eastern hemlock. In the Great Smoky Mountains, instead of the scarlet color of sugar maple leaves and the gold foliage of beech filling coves near the national park's Sugarlands Visitor Center, we can expect to see ever expanding leafy masses containing the tawny-port bronze of flowering dogwood, the purple-tinged brown leaves of those oaks remaining after the gypsy moth blight, the lemon yellow of hickories, and the bold green needles of southern pines, all of which will be able to tolerate the warmer and drier climate of the future. Middle and upper slopes on the mountains should provide a refuge for most of the remaining species of hardwood trees now living in the Great Smokies. The highest elevations will be havens for mixed forest communities of beech, blue beech, red maple, mountain ash, black cherry, eastern hemlock, and possibly sassafras. The loss of the sugar maple's intense reds from the autumn canopy will be compensated by the addition of a leafy kaleidoscope of purples, reds, oranges, and pinks from black gum, sweet gum, sourwood, and red maple trees.[40]

The upside for leaf peepers is that, even though the species mix will change, the full spectrum of fall colors will survive. Hardwood forests can migrate up the Great Smoky Mountains more than a mile in altitude before they run out of room on the high peaks. The fall foliage displays will persist; they will just be perched higher near the mountain summits. The successful Greenhouse-world strategy for many forest species will be to migrate along with suitable climate regimes toward higher elevations, or farther north into New England along the corridor of the Appalachian Mountains.

Elusive Ecotones

If forest species spread upward from low elevations, what will happen to the vegetation *now* clothing the mountain summits? What will happen, that is, to the position of the **ecotones**, ecological zones of transition that form the altitudinal boundaries between major types of montane vegetation? Peter White, conservation biologist and director of the Botanical Garden, University of North Carolina at Chapel Hill, has studied the rare plants that are today restricted to high elevations within the southern Appalachian Mountains. It turns out that of the truly rare and endangered species of plants living in the Great Smoky Mountains, nearly 30 percent are found nowhere else on the earth but in the Great Smoky Mountains. These rare plants grow in special and precarious places, clinging to craggy outcrops of bedrock, exposed to the ravages of fierce wind, sleet and ice, and recurring landslides. Another 40 percent of the rare plants found at high elevations also grow farther north along the Appalachian Trail in the mountains of New England. An additional 20 percent reside in spruce-fir forests or grassy "balds" and are relics of the Pleistocene, a time when a colder climate ecologically connected the high peaks of the southern Appalachians with the Canadian Arctic. These rare plants are **disjunct in distribution**, that is, they are remnants of populations that were more widespread during the Pleistocene. In some cases they have been genetically isolated in islandlike montane habitats so long that they have become distinctive new species and are called **endemics**.[41]

Using clues from plant fossils preserved in mountain bogs, we have probed the distant past to find a key for projecting the future of these threatened species.[42] To envision their fate in the Greenhouse world, we have to go back to the Pleistocene, to an ice-age world that last existed some 13,000 years ago. At that time, the summits of the Great Smoky Mountains were covered with snow and ice most of the year. Patches of snow-free soil were colonized by cold-requiring, herbaceous alpine plants. Today, these alpine herbs grow on mountain peaks in the northern Appalachians and in the grassy tundra vegetation of the high Arctic regions of Canada. When climate changed from ice-age to ice-free, about 10,000 years ago, alpine tundra was replaced above 4,500 feet elevation in the Great Smoky Mountains by invading spruce and fir. If our highest peak, Mount LeConte, were only another 500 feet higher, a climate-controlled **treeline** would still exist in the Smokies, with alpine tundra meadows growing on the summit and evergreen trees marking the transition zone, or vegetation ecotone, between tundra grassland and evergreen spruce-fir forest.

Along with tundra's response to warming climates of the past, boreal evergreen forests underwent major adjustments to the end of the last ice age. Cold-loving trees such as spruce and fir died out across much of the southeastern United States and gradually became confined to their present mountaintop homes above 5,500 feet elevation, the ecotonal border between cold-loving evergreen conifer forest and warmth-requiring temperate deciduous forest. This ecotone is anchored by a bioclimatic threshold, an average July temperature of 63°F, the warmest climatic limit tolerated by spruce-fir forest.[43]

Now, back to the future. The fate of nine out of every ten species of rare plants in the Great Smoky Mountains will be determined by the rate and amount of Greenhouse-world climate warming that takes place *in the next few decades*. With a doubling of atmospheric carbon dioxide by the Boomer Breakpoint year 2070, the GISS model forecasts a 5°F increase in average July temperature across eastern North America. Hikers understand that summer temperatures cool off about 1°F for each 300 feet of elevation they climb up a mountainside. A 5°F increase in midsummer temperature means a rise in the critical threshold for spruce-fir forest survival of some 1,600 feet, which unfortunately would push their range above the highest peaks in the southern Appalachian Mountains. The southernmost limit of these mountain ecotones will retreat northward by some 120 miles. We think that some enclaves of spruce and fir forest might survive climate warming in small "pocket wildernesses" with suitable soils and microclimates, especially in high-elevation saddles near mountain crests that receive cold air drainage and thus stay cooler than the overall surroundings.

The GISS projections are conservative. Their predictions of massive biotic change, but not wholesale extinction, are barely tolerable to conservationists. Should the dire projections of the GFDL model prove more accurate, however, summer warming could increase by as much as 12°F. Such bioclimatic stress would cause extinction of *all* arctic-alpine and boreal plants from southern latitudes of the Great Smoky Mountains. The amount of Greenhouse-world warming envisioned in the GFDL scenario would collapse the tundra and boreal forest communities. These montane ecotones would be displaced up the slope by 3,350 feet, which means they would run out of mountain slope to climb! In their northward flight, these endangered communities may be forced out *even from the northern Appalachian Mountains*.[44]

Adapting to a Future Mountain Scenario

What do altitudinal changes in ecotones mean to Boomers? We can use the analogy of a championship baseball game. We Boomers are in the bottom of the ninth inning, with two runners on base and two outs. The score is one to nothing, and we're behind. What's at stake? Not merely the survival of the spruce-fir forest but also our endangered vistas, the quality of clean air, pure water, and the solitude we seek in the mountain wilderness.

Here comes the first pitch. Acid rain and high ozone concentrations damage vulnerable red spruce trees. *Strike one!* Now the second pitch. Fraser fir trees that have been under insect attack since the 1950s by the European balsam wooly adelgid die back to one-tenth of their former numbers. *Strike two!* Third pitch: Greenhouse warming intensifies. *Strike three!* The spruce-fir forest is out. It will be extinct in the southern Appalachian mountains by the Boomer Breakpoint year 2025, and possibly as soon as the Boomer Breakpoint year 2010.

In the Greenhouse world, in natural biological communities as well as in human society, there are going to be winners and also losers. Win or lose, the important question now is, Who can adapt? Ever increasing temperatures will shift the seasonal cycle into an optimal human comfort zone during some or much of the year, depending on the latitude at which we live. We will lose some native plants and animals to extinction, however, and many of those species are already threatened by the overuse of land in their natural ranges, which has restricted them to ever smaller sanctuaries such as the Great Smoky Mountains. In the twenty-first century, humans must learn to adjust to change as well as to find ways to help our fellow creatures avoid extinction. In adapting to a twenty-first-century mountain lifestyle, we will have to look in new places and at different times of the year to get the most from our experiences as wildflower pilgrims and leaf peepers. And we'll have to tread lightly to make sure we don't further damage sensitive ecosystems in pursuit of our chosen leisure activities.

CHAPTER

7

The Sunbelt

s newspaper editor of the prestigious *New York Tribune*, Horace
Greeley advocated in the mid-1800s that Americans move to
the New Land of Opportunity. His advice was "Go west,
young man." Perspectives have reoriented about ninety
degrees. As recent trends make plain, some inner voice must now be telling
us, "Go south, Boomers"—south to the sunbelt. As we begin to migrate *en
masse,* will we find those fabled cities of gold with endless summers? Will
we discover our superoasis in a time of change, or see it changed to a
superdesert in our lifetimes? Will we simply find relief from slushy snow
and bone-chilling winters, and maybe even drop a stroke or two off our golf
games? Will our particular choice of sunbelt location, in the company of
many thousands or even millions of fellow migrants, render us vulnerable to
ecological insecurity in the form of electrical brownouts, fire ants on the
golf course, and Asian termites chomping away the walls of our retirement
homes? Will we leisurely tend idyllic Mediterranean-style herb gardens to
garnish our gourmet dinner tables, or will we be spending our spare time
weeding out pesky alien invaders encountered among the plants in our
backyards? Will we be caught up in the next Boomer wave as ecological
refugees, fleeing from one apparently safe site to the next in a deteriorating
Greenhouse world, or can we truly find our place in the sun?

In this chapter, we consider a host of problems that will arise in a
Greenhouse world, from enhanced variability in weather patterns to water

shortages, dust-bowl droughts, less predictable energy supplies, and pervasive disruption of native habitats. We draw examples primarily from the **sunbelt** region, stretching from the southeastern United States through the high plains of Texas and Oklahoma across to the desert Southwest of New Mexico, Arizona, Nevada, and California. Many of the dilemmas that retiring Boomers will face in sunbelt locations will also affect the quality of life elsewhere.

Ecological Refugees

Recently, we met one of our fellow ecology professors in the departmental office at the University of Tennessee. Just back from fieldwork in the high plains of central Texas, Gary McCracken was clearly agitated. "There's water running in the streets," he exclaimed, "and they're killing my bats!" He was referring specifically to Mexican free-tailed bats, and in Texas millions of them roost in caves, sheltered back in the hill country.[1] On evening flights in their quest for food, these bats search out swarms of insects such as corn borer moths, major agricultural pests whose larvae devour vast fields of corn and cotton.

The connection between water, corn borers, and free-tailed bats is a long story. On the arid Texas plains, production-oriented farmers mine groundwater to irrigate their crops. They have learned well—perhaps too well—the harsh lessons taught by the Dust Bowl era of the 1930s. By 1940, extensive irrigation projects were tapping the liquid gold that is groundwater. Now the supply of precious water is plentiful and predictable, and the attitude of some farmers and ranchers to its use is cavalier. Our colleague was livid about the unnecessary waste of well water being spilled onto the streets as overflow from pumps.

Thirty percent of all the freshwater used for irrigation in America is pumped from one source, a vast natural reservoir in the Southern High Plains called the **Ogallala Aquifer**.[2] It is essentially a deep, slow-moving underground river—a subterranean layer of water-saturated sand and gravel some 150 to 330 feet thick. The top of the gravel generally lies only 50 to 330 feet below ground. Ogallala water is rainwater that fell onto hill country and runoff from Rocky Mountain glaciers that sank underground ten thousand or more years ago. This Pleistocene-age groundwater is a *non-*

renewable resource, since rainfall in modern times has been too sparse to resupply the underground Ogallala River and keep it flowing. At present, water is being pumped out of the aquifer much, much faster than it can be recharged. During the late twentieth century, groundwater levels in parts of Texas, Oklahoma, and Kansas have dropped more than 100 feet because of the extensive irrigation of corn and cotton. The underground river is drying up.[3]

Farmers in some areas are beginning to conserve water and plant drought-tolerant crops. But, for Gary McCracken and his bats, this is even worse than letting the water run in the streets. Without food provided by such crops as corn and cotton, corn borers will become less common. No insect pests means no more free meals for millions of free-tailed bats. Even without the further complications of worsening drought brought on by Greenhouse warming, the bats, which are already quite restricted in distribution, face a perilous future. They must become **ecological refugees** and move to new homes to survive.

Similarly, farming is threatened throughout the Southern High Plains, both because of loss of Ogallala groundwater and because of impending summer droughts envisioned for the sunbelt in the Greenhouse world.[4] Even so, agriculture in America's heartland is still generally governed by the *economic* imperative to grow more crops, not by the *ecological* imperative to conserve groundwater. Some hydrologists say that we have only twenty years worth of deep groundwater left in the Ogallala Aquifer. They predict that the extent of irrigated land for crops and cattle will diminish by 80 percent across the central high plains by the year 2020. This is not an isolated or regional problem. In over half of the United States, groundwater levels are dropping because, each year, more than one-quarter of the amount pumped out is not being replenished.

Someday, a novelist will write the Boomer sequel to John Steinbeck's *The Grapes of Wrath,* a chronicle of the next great dust bowl with Americans forced from the expanding man-made desert of the Southern High Plains.[5] (What was the title of that song? Something like "Happiness is seeing Lubbock, Texas, in my rear-view mirror"?) Within the next few decades, both bats *and* Boomers may well become ecological refugees.

Water Wars

Three water specialists, Wayne Solley, Robert Pierce, and Howard Perlman, monitor the pulse of long-term water use in America. In a 1993 report, they take the long view:

> It seems likely that water withdrawals for public supply and domestic uses will continue to increase as population increases. . . . With increased competition for water and for instream uses, such as river-based recreation, aesthetic enjoyment, fish and wildlife habitat, and hydroelectric power, along with higher municipal uses, irrigators will have increasing difficulty competing economically for available water supplies. Municipal and industrial users can afford to pay much more for water than the farmers.[6]

Is this multiple *use,* or is it multiple *competition* for water?

The stage is set for water wars. In the hill country of west Texas, oilman and entrepreneur T. Boone Pickens is adapting the proven technology of drilling and transporting petroleum in order to seize a new opportunity. Rather than pump the "black gold" of oil, Pickens plans to pump another liquid gold, freshwater. Organizing a consortium of ranchers and farmers, Pickens will literally tap into rural groundwater reserves and use a pipeline network to carry liquid gold to high-paying customers, the water-parched municipalities whose continued growth is constrained by shortfalls in water supplies. T. Boone Pickens's new company, Mesa Water, Inc., has already secured water rights to a volume of more than 100,000 acre-feet of groundwater, with five delivery routes planned to serve San Antonio, El Paso, Dallas, and Fort Worth. For more on the emerging market for this "futures" commodity of freshwater, read Kenneth Frederick's national overview prepared for the Washington, D.C., think tank Resources for the Future.[7]

It's a push-pull situation for Boomers. We're being *pushed* from localities suffering from deteriorating water quality, expensive and unreliable water supplies, and increasingly intense municipal oversight of our personal use of this precious water. We are *pulled* toward attractive recreational areas with clean, plentiful water in rivers and dammed reservoirs, which also are a powerful magnet for utilities and industry. The drawing card is cheap electrical power.

Major producers of electric power (aside from hydroelectric) require vast quantities of cooling water in order to generate electricity from fossil fuels, nuclear materials, or geothermal sources. These industrial facilities are by far America's largest users of water, with irrigation coming in a distant second place. Today the big sunbelt producers of electric power are located in California and in states bordering the southern Atlantic Ocean and the Gulf of Mexico. Over 99 percent of their needs are satisfied by surface water (two-thirds freshwater, one-third saltwater), most of which is used to cool steam in condenser towers and nuclear reactors. Some of this cooling water is lost by evaporation; the remaining hot water is returned to collecting ponds, rivers, and lakes.[8]

How will our future supply of electric power be affected by Greenhouse-world water resources? The potential for conflict can be seen readily in one popular sunbelt destination, the **Tennessee River watershed,** which covers 40,900 square miles. The water resource region of the Tennessee River covers portions of seven states: Tennessee, Virginia, North Carolina, Georgia, Alabama, Mississippi, and Kentucky. Based in Knoxville, Tennessee, the Tennessee Valley Authority (TVA), a federal agency, produces electric power for the region. TVA provides integrated management control over coal-fired and nuclear-powered plants and hydroelectric dams. In conjunction with eleven other local public power companies, the Knoxville Utilities Board (KUB), and local environmental groups, TVA has implemented a valleywide pilot program called *Green Power Switch* to help ease the regional dependence on nonrenewable energy resources.[9] For an added price to monthly utility bills, TVA customers will be able by the year 2003 to choose more expensive but environmentally friendly alternatives for cleaner, greener electricity. This **"green power"** is generated from renewable resources such as the water flow of TVA reservoirs, banks of solar-powered photoelectric cells, wind-powered turbines, and landfill-generated gas. In the initial market test, eight megawatts of green power will provide 150 kilowatt hours a month to 30,000 homes and businesses. As the newsletter *KUB Connection* beckons, "So, join us and plug in to green power—the energy that saves tomorrow!"

In addition to producing electric and hydroelectric power, the Tennessee River serves many needs. TVA is responsible for flood control and the recreational use of forty-two major reservoirs, and for navigational access along nearly seven hundred miles of the river for commercial barge traffic and pleasure watercraft. TVA also monitors water quality and aquatic resources to ensure suitable drinking water supplies for riverside municipalities such as Knoxville and Chattanooga. TVA's service area for supplying electrical power is nearly twice the size of the watershed basin. This is one very important watershed.

How vulnerable is the Tennessee River Basin to changing water supply and demand? We can gauge this in several ways. Hydrologists calculate **relative storage capacity** as one measure of a region's ability to withstand extended intervals of flood or drought.[10] This measure is the ratio between the maximum volume of water that can be stored in the drainage basin and the typical rate of renewable supply, in other words the amount of water a basin can hold compared to the average annual input of new water. For the Tennessee River, the current value of 0.23 means that TVA can capture and store less than one-fourth of the annual flow. In other words, it still has too little storage capacity in its reservoir network to protect against big floods or to buffer long-term water shortages.[11]

Another way to look at this problem is to compare consumer demand with the average annual supply of renewable water coming into the drainage basin. Total demand is the sum of water used by consumers, transferred to other places, lost through evaporation, or pumped as nonrenewable groundwater. For the Tennessee Valley, this **relative demand** is only 0.01; that is, people use only one percent of the water that flows into the basin each year. This means that future economic development should not be limited by lack of water in the TVA region.

El Niño, again and again and . . .

What should Boomers expect ahead in the Tennessee River Valley? Hydrologists are preparing for increased variability in annual and decadal weather patterns. Higher temperatures will mean greater losses of water by evaporation, with the consequence that more water will be recycled across the watershed through higher amount of rainfall. Even small changes in evaporation and rainfall can result in major changes in stream runoff and available water resources.[12]

An important part of the increasing weather variability in the U.S. sunbelt may be linked to changing temperatures in the tropics, particularly in the eastern Pacific Ocean. Surface waters there sometimes warm dramatically and flow toward South America by late December, producing a distinctive kind of ocean current, air circulation, and global weather pattern called **El Niño** (translated "the Christ child"). In the following winter, the second weather child, **La Niña** ("little girl") sets up a different global weather pattern as the ocean surface cools. The alternating El Niño/La Niña weather oscillation is one of the major sources of global climate variation. In the future, Greenhouse global warming may shorten the time between occurrences of El Niño/La Niña cycles, currently every three to seven years. As Curt Suplee reported in *National Geographic* magazine, "There is a consensus among climate scientists that El Niños have become more frequent and progressively warmer over the past century."[13]

Climatologists such as Bob Henson and Kevin Trenberth have speculated that the near future holds the prospect for even more frequent and severe El Niño and La Niña events.[14] Henson and Trenberth note that state-of-the-art computer models can now simulate these powerful ocean-atmosphere weather patterns for Greenhouse-world conditions. They suggest that regional cycles of rainfall torrents alternating with severe drought may be accentuated by global warming. If clustered back to back over a number of consecutive cycles, El Niño and La Niña could become the predominant weather influence all across the American sunbelt, producing alternating warm-wet and warm-dry conditions.[15] Such an increasingly powerful cycle would profoundly alter the flow of water through watersheds.

Thirsty, hot, and sitting in the dark

Barbara Miller and W. Gary Brock, based on their research at TVA's Engineering Laboratory, have projected how the Tennessee River is expected to respond in a Greenhouse environment.[16] Miller and Brock started with the GISS scenario envisioned for the Tennessee Valley region *as soon as the year 2030.* That Greenhouse-world forecast indicates that average annual temperature should increase by 8°F over today's average. Despite a small reduction in cloud cover, the warmer, humid air should hold about one-third more water vapor. Yearly precipitation levels should be slightly greater, by about 2 percent, than modern values. Year to year, however, total precipitation may fluctuate widely, from one-third less to one-third more than today. Therefore, the total volume of runoff that will be funneled through the

Tennessee River is calculated to average about the same as today's. Yet it may become highly variable, from one year to the next, because of alternating strong El Niño/La Niña years. Rivers tend to magnify rainfall effects. During flood years, *twice* as much water may flow down the Tennessee River as it now carries during a typical year. Drought years may reduce the flow to as little as *one-fourth* of today's average (figure 7-1).

GISS projections are conservative, remember, giving *minimum* estimates of the changes to be expected. In the extreme GFDL simulation of hot, very dry weather, seasonal droughts would be predicted to cut off water supply altogether to the Tennessee River, thus producing *no runoff* for up to seven months each year! Such GFDL simulations were considered unthinkable by TVA researchers, who chose instead to focus on possibilities envisioned by the less extreme GISS model.[17]

The Greenhouse-world climate simulated by the GISS model includes more frequent storms with intense rainfall. For example, in the warm-wet GISS scenario (what we call the El Niño scenario), the peak storm season will occur in late winter and spring (figure 7-1). These pulses of precipitation will amplify the severity of March floods and potentially threaten the structural integrity of the TVA dams. *The dams may fail.* Predicted flood levels exceeding fifty feet above modern flood stage would not be contained by the banks of the Tennessee River. *These Greenhouse-world floods would be more severe than any historic floods* and would be capable of inundating major sectors of downtown Chattanooga and Knoxville. You may want to factor in "elevation above Greenhouse-world flood zone" when you select a lakeside cottage along a sunbelt reservoir.

Even the traditionally dry seasons of summer and fall may become increasingly prone to catastrophic deluge in the southeastern sunbelt. More frequent and greater-magnitude hurricanes will trek northeastward from their spawning waters in the Gulf of Mexico. After making landfall, even "downgraded" tropical depressions will cause flooding. They tend to stall out as they encounter the blocking southern Appalachian Mountains. This mountain barrier will catch torrential rainfall, dumping peak runoff into the headwaters of tributary streams and eventually into downstream reaches of the Tennessee River.

In their "first-cut assessment," Miller and Brock considered the watershed impact of these Greenhouse-world shifts. The warm-wet GISS scenario

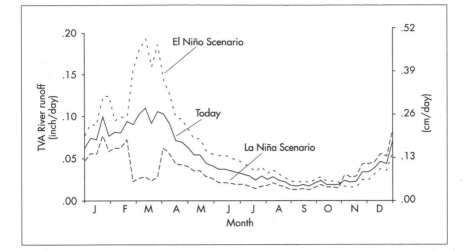

Figure 7-1. **TVA projections of annual river runoff.** During the interval of Boomer Breakpoint years 2030–2070, alternating warm-wet (El Niño) and warm-dry (La Niña) rainy seasons will cause stream levels and river flow to fluctuate wildly above and below today's "averages." Expect major flooding but also severe droughts in the Tennessee Valley region. (From Barbara A. Miller and W. Gary Brock, *Sensitivity of the Tennessee Valley Authority Reservoir System to Global Climate Change,* Report no. WR28-1-680-101 [Norris, Tenn.: TVA Engineering Laboratory, 1988].)

(our El Niño scenario) projects only a one-third increase in *average* yearly rainfall, which would enhance stream flow and permit more water to be stored in TVA reservoirs from winter through summer (figure 7-1). Under these conditions, TVA should be able to meet its minimum water-release requirements for navigation and yet maintain water levels in the reservoirs at least ten feet above present levels for most of each year. Reservoirs filled to capacity would benefit recreation and sportfishing, and they should still allow 16 percent more power to be generated from TVA hydroelectric plants (figure 7-2). The probability of flooding would increase, however. Water levels may have to be kept lower seasonally to ensure that reservoirs have enough storage capacity for flood protection downstream.

Yet the worst may still happen. A negative outcome of the GISS scenario is that extreme floods may damage Boomers' riverside homes as well as highway bridges, railway trestles, docking facilities for navigation, and municipal treatment plants for drinking water and for sewage. Even

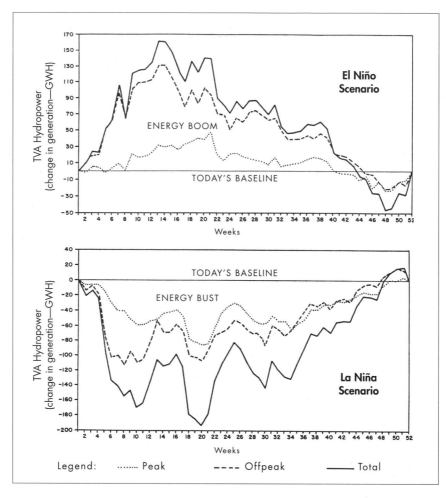

Figure 7-2. **TVA projections of annual hydropower production.** During the interval of Boomer Breakpoint years 2030–2070, alternating warm-wet (El Niño) and warm-dry (La Niña) rainy seasons will cause today's energy booms and busts to intensify. This could lead to longer and more frequent electrical power brownouts and blackouts, courtesy of global warming. (From Miller and Brock, *Sensitivity of the Tennessee Valley Authority Reservoir System to Global Climate Change*, see fig. 7-1.)

without floods, the "normal" rainfall of wet years would mean increased erosion. Sediment carried by the river during times of peak runoff may change the shape and depth of the river channel. Flood-prone riverways would require extensive dredging and stabilization of eroding river banks. In the Greenhouse world, it may become more difficult for commercial barge traffic to travel up the Mississippi, Ohio, and Tennessee Rivers as far as the inland port of Knoxville.

Dry years pose a different set of problems. The warm-dry GISS model of Miller and Brock (what we call the La Niña scenario) would reduce average annual rainfall by nearly one-third, reduce stream flow, and endanger drinking water supplies for forty-nine municipalities (figure 7-1). This would cause water shortages for the increasing Boomer populations initially enticed to the sunbelt by the prospects of sportfishing and sailing on man-made TVA lakes. During warm and dry years, the reservoir network would store substantially less water, disrupting navigation and cutting production of hydroelectric power by one-fourth (figure 7-2). Shortfalls in energy production would coincide with peak summer demands for electricity to run air conditioners. As dropping water levels began to expose extensive mud flats, algal blooms would become rampant in the warm, muddy water, and sport fish would suffer massive die-offs. Boomers, as well as subsequent generations, would eventually shun the once-attractive "Great Lakes of the South."

In follow-up studies, Barbara Miller and colleagues spelled out the Greenhouse-world implications for TVA's ability to generate electricity.[18] The TVA power-producing network includes coal-fueled plants, nuclear plants, and hydroelectric dams. Heat produced by fossil or nuclear fuels converts river water into steam, and steam-driven turbines produce electricity. The steam condenses back into liquid in cooling towers, and then this hot water is cycled through collecting ponds back to the reservoir. During El Niño years, consistent with the warm-wet GISS model, high precipitation would sustain high reservoir levels and increase the production of hydroelectric power by TVA (figures 7-1 and 7-2). During years dominated by the La Niña weather pattern, as simulated by the warm-dry climate model, reduced stream flow and shrinking reservoirs would all limit the amount of water available to the TVA hydroelectric plants. But nuclear and coal-fueled plants would feel the pinch as well. If the cooling water in reservoirs becomes warmer than 82°F, federal regulatory agencies shut down the plant's water intake. Hot water reduces both the efficiency and amount of power produced by steam-driven turbines. Even if the plants' intake remains open, however, it may not be able to operate. EPA holds the last word on the "thermal discharge." Condensed water must cool *below* 97°F before it can be returned legally to the reservoirs.

La Niña conditions are the weak link in TVA's energy future. Hot-dry summers in the South will disrupt the predictable production of electricity from all TVA plants. Federally imposed shutdowns would be regular occurrences during the worst hot and dry years in a Greenhouse world. As power plants drop offline, Boomers will experience the reality firsthand. Imagine the disruption and discomfort of flickering daily brownouts or even blackouts as the energy net regularly goes down. TVA already refers to such power shortages, not yet frequent, as "operational headaches," or "inconveniences in meeting system load requirements during the peak [summertime] demand periods."[19]

The rational solution for communities across the Tennessee Valley is to actively support TVA's innovative pilot program. Responsible individuals, companies, and municipal governments can all participate in the *Green Power Switch* to broaden the output of dependable power from renewable energy sources.[20] TVA offers cash rebates for new home construction that incorporates energy-conserving architectural solutions, such as energy-efficient windows, adequate insulation, heat-pump technology, and a passive solar cooling design. If you want to live off the grid of electrical power lines, consider installing photovoltaic cells. A solar power system can be a very effective individual solution in this region of the sunbelt. But demand might still outrun supply. Increasingly, individual choice in lifestyle alternatives can dramatically reduce the overall *demand* for electricity, both on and off the grid. Many Boomers are adopting ingredients of a simpler, more self-sufficient lifeway practiced by Amish and Mennonite farming communities. To explore the possibilities for a traditional nonelectric homestead, see the catalog of appliances and gadgets in the *Good Neighbor Heritage Non-Electric Catalog* from the Amish-country Lehman's Hardware and Appliances store, Kidron, Ohio.[21] Chelsea Green publishes many books about self-reliance and sustainable living; see the last page of this book for more information.

Does it seem like too much work to plan ahead constructively? Consider your alternative! You could be caught in a peak-summer blackout of the power grid. What can a sweating Boomer do? For a short-term fix, you might buy a gasoline-powered electrical generator, install it in your home, and learn how to use it. Better yet, you can invest in the stock of companies that make these emergency generators. While you're waiting for that to pay off, you might reconsider planning ahead. Your immediate possibilities are limited—move it, adapt creatively, or just plain swelter! You might adopt a seasonal backup plan to escape the peak summer heat by going to your time-share condo or cabin in a cooler setting in the mountains or at a northern lakeshore. If becoming an ecological nomad doesn't fit your contemplated future lifestyle, you may wish to retrofit your home. Next on your list of things to do, buy that flood insurance while FEMA is still offering it—flooding does happen in river flood plains! If none of this seems worth the bother, then prepare yourself as best you can for the time when blackouts wink off your electricity, leaving you thirsty, hot, and sitting in the dark.

Desert oasis

Greenhouse-world projections indicate warmer surface temperatures in the eastern Pacific Ocean, reinforcing El Niño weather patterns. *Prudent risk management* means planning for the increased intensity and frequency of El Niño events. How will they impact the western United States?

Pacific coastal storm tracks are expected to shift southward, focusing greater winter rainfall particularly in California, Nevada, and Arizona. Count on a reduction in winter snowpack on the summits of the Sierra Nevada and the Rocky Mountains; the winter snow season will shorten by one or more months. Also, expect an earlier peak runoff with snowmelt arriving in a more rapid spring pulse. Summer streamflow will decrease markedly. By the year 2090, California's climate is projected to warm by 5°F, with a doubling in winter rainfall; the average annual runoff from streams should also double. But these so-called average conditions are a statistical fluke—the Greenhouse-world reality will be alternating years of intensive floods and severe droughts. More extreme wet years will be offset by more extreme dry years. This heightened seasonal contrast between winter-wet and summer-dry conditions will make freshwater availability less predictable in the future.[22]

We can easily predict, however, that more people will be *needing* water. Since 1950, regional populations in the American Southwest have quadrupled, with most people now living in cities. Southwesterners, numbering 48 million in 1999, are expected to expand to between 60 and 74 million by Boomer Breakpoint 2025, presuming a demographic scenario of only modest growth.[23] In California, where substantial population growth is anticipated, urbanization increasingly will spill over onto adjacent agricultural lands, according to the *California Water Plan*, which projects the needs and options for water resources for the next twenty-five years.[24] A doubling of rainfall, with accelerated stream runoff, will trigger more mudslides and flooding, which will destroy cities and suburbs built on floodplains. Outside the cities, semiarid grassland and shrubland will be replaced by woodlands and forests.

Today, California residents use freshwater pumped from the Colorado River and from its northern state reservoirs. Over the past ten years, California has consumed more than its legal, normal apportionment of Colorado River water.[25] Extending from the eastern Sierra Nevada, the watershed of the **Lower Colorado River Basin** covers most of Arizona, Nevada, and parts of southern California. The river headwaters of the Upper Colorado Basin encompass Wyoming, Utah, Colorado, New Mexico, and some of Arizona. With many states sharing this vital freshwater lifeline, formal agreements (based on historic wet years with abundant river flow) have carved up the proportion of river water granted to each state. Today, 96 percent of the renewable water supply to the Lower Colorado River Basin *is used each year*, leaving precious little flexibility to accommodate increasing demand or adjust to future Greenhouse-world climate variability.[26]

A 1993 EPA Report, *The Colorado River Basin and Climate Change*, warned that the entire 243,000-square-mile watershed is "extremely sensitive to climatic changes that could occur over the next several decades."[27] Plausible scenarios predict a greater variability in river flow, oscillating between 30 percent higher and lower than today's levels. Even modest assumptions about near-future climate change all foretell potential problems, with episodic late-winter flooding, less water in dammed reservoirs, hotter, drier summers with greatly increased evaporation, and rising salinity in the no longer fresh water of the Colorado River. Overall, higher variability in precipitation will degrade the quality of water supplied by the Colorado River. Water users in southern California, Nevada, and Arizona are at high risk for thirsty times ahead.

Trouble in paradise?

California dreamers face long daily commutes and traffic congestion, car exhaust and degrading air quality, high property taxes, and the troubling scarcity of clean, germ-free water. Demographers project an out-migration of Boomer refugees from California to Nevada. Some are leaving already, capturing substantial capital gains by selling expensive homes and escaping traffic-clogged expressways, ozone alerts, and asthma attacks for the fresh air and sunny skies of the wild blue yonder. Where are they going? Click on the *Money* magazine Web site to sort through an extensive database of nearly five hundred towns Americans are choosing today for retirement.[28]

The fastest-growing large city in America has become the top retirement destination. With 300 sunny days each year, Henderson, at the southern tip of Nevada, gives the term "sunbelt" a good name! How fast is fast-growing? In 1980, Henderson boasted a population of 24,376. In 1990, it was up to 67,127, and in 2000, 189,000 people call Henderson home. City planners project that their population will reach 560,000 in the next several decades.[29] Henderson, once considered a suburb of Las Vegas, is now Nevada's second-largest city. Aggressive electronic marketing on Henderson's Web banner heralds "the fastest growing city on the Internet," with concierge service and logistic support welcoming new arrivals. Henderson's municipal Web site serves as the city's virtual control center, using the Internet to update the city map, issue directory names of new streets *every few weeks*, track new construction developments, and electronically post future listings for master neighborhood plans. Henderson's Redevelopment Agency oversees construction of municipal infrastructure and coordinates service programs, in spite of the obvious strains imposed by the massive building boom. Some 96 percent of current residents are satisfied with the existing quality of life; interpreted as a positive spinoff of Henderson's development, airborne dust swirling upward from construction sites enhances the spectacular sunset colors in the sky at dusk.

As Rob Turner reported for *Money* magazine, the "city goes out of its way to maintain a utopian ideal in the face of overwhelming growth." The weak link in this city's planning is water. The Southern Nevada Water Authority (SNWA) is building additional water pipelines, pumping stations, and even more reservoir capacity for collecting freshwater. But these efforts to augment water needs, and this infrastructure plan, are playing catch-up *at best!* Expected to be completed by 2008, SNWA's new regional water treatment facilities will deliver to Henderson and the Greater Las Vegas area freshwater from the Lake Mead reservoir. The ironic problem here is that

Lake Mead is filled by water *flowing from the Colorado River*. To quote Rob Turner, "For now, Mayor [James] Gibson believes the town has struck a balance between economic growth and quality-of-life demands." At its current rate of expansion, Henderson's Mayor Gibson expects to face a water crisis within the next ten to twenty years.[30]

Henderson's Boomer influx has overextended its infrastructure capacity. In the Greenhouse world, shortfalls of Colorado River flow will make the water supply ecologically vulnerable by the *first* Boomer Breakpoint in the year 2010. That's when the *real* Boomer influx will be just getting started (figure 2-2).

Birds, the Consummate Ecological Nomads

Hard-working Boomers watch with jealous anticipation the seasonal odyssey of **snowbirds**, those retirees who flit back and forth, footloose and fancy-free, between rustbelt havens and sunbelt refuges. Seasonal snowbirds track their avian namesakes, alternately seeking optimal comfort zones for winter warmth and summer coolness. Clutching binoculars and Audubon field guides, birders make a lifestyle of their passion for following and hopefully glimpsing elusive songbirds. The South attracts bird lovers anxious to add new feathered species to their "life lists."

The most comprehensive database for estimates of the abundance of bird species across North America is maintained by the National Audubon Society. Since the year 1900, bird watchers have made a tradition of counting America's bird populations each Christmas. Each local bird count covers the area of a circle fifteen miles in diameter. Each year more than 45,000 people participate in the census in nearly 1,700 locations across North, Central, and South America as well as the Caribbean and Pacific islands. The Christmas bird counts give a basis for understanding the ecology and seasonal migration patterns of many common and rare bird species.[31]

Ornithologist Terry Root has studied the big picture of bird migrations by pulling together the bird-count data in a new way.[32] Root evaluated the data for 148 species and subspecies of birds commonly seen during daytime over land, using a decade of observations from 1962 to 1971 made at 317 Christmas count sites across Canada and the United States. Four distinctly different groups of land birds were included, each with specialized feeding and habitat requirements. Raptors and shrikes included 12 species of relatively large-bodied, free-flying predators whose food supply consists of smaller birds, mammals, and large insects. Bark gleaners included 20 species of woodpeckers, nuthatches, brown creepers, and black-and-white warblers who harvest insects hiding within irregular cracks of shaggy bark on deciduous trees. Foliage gleaners—25 species of warblers, wrens, kinglets, gnatcatchers, and vireos—track habitats where their prey of insects and spiders are easily detectable and abundant. The fourth guild of birds, the seed eaters, consisted of 31 species of sparrows and finches, who occupy tall grasslands and dense brush, habitats offering both protective cover and plentiful seeds.

Root compiled the bird counts into an atlas of maps showing the winter patterns of abundance and distribution for the species.[33] By comparing the winter range boundaries of individual species with environmental variables, Root identified key ecological factors including four climatic thresholds and two landscape features that control the whereabouts of the species. Lowest January temperature, amount of yearly precipitation, humidity, and length of frost-free period, as well as elevation and type of vegetation, were all found to be important in determining where species of birds live today.

Birds seeking their winter home are drawn to a milder climate, without extreme cold, and to a landscape providing the right kinds of shelter and food for overwintering. Larger birds are seen farther north in winter than smaller birds because, with their larger body mass, the big predators can better tolerate cold weather. Even raptors, however, opt for the warmer sunbelt, where they encounter the highest abundance of small-bodied birds to prey upon. Raptors thus tend to move out of summer breeding areas, which become harsh and hostile habitats in winter, following their winter food supply south.

Smaller bark gleaners overwinter in temperate deciduous forests of the eastern United States, and foliage gleaners concentrate in the Deep South, where their prey of insects and spiders remain abundant and active all winter. Seed gleaners are drawn to the kinds of southern vegetation that ensure both an abundant supply of seeds and shelter in which to hide. Many smaller species of birds are attracted to the Atlantic and Gulf shores, where steady ocean temperatures moderate a mild coastal climate with a long frost-free period.[34]

How will birds readjust their territories in response to changing Greenhouse-world habitats? Terry Root and climatologist Stephen Schneider have explored the nature of the **first alert** we should expect from bird communities. Some birds, they noted, will be able to change their ranges as rapidly as the Greenhouse climate changes. Other birds may not be able to shift as far or as fast for any number of reasons, including their own physical limits, complete loss of habitats, or attempts to cash in on new opportunities for food and shelter in the changing environment. Still others will remain comfortable where they are. "Such differential movements of species," Root and Schneider predicted, "will certainly cause a tearing apart of communities, thereby forcing potentially dramatic restructurings and reorganizations." The scientific findings of Root and Schneider specifically address the ecological consequences of global warming:

> This implies that substantial disequilibrium within ecosystems could be created owing to maladaptations [of species to their climatically changed habitats], significant shifts in species ranges, and inevitable extinctions. Consequently, the only outcome that can be predicted with virtual certainty is major surprises. The only forecast that seems certain is that *the more rapidly the climate changes the higher the probability of substantial disruption and surprise within natural systems.* Dramatic disruption of communities can be expected to occur in the next [twenty-first] century.[35]

The EPA *Global Warming* Web site monitors the climate changes that will potentially cause disruption within natural communities. The first alert from birds has already sounded.[36] Some species of birds have been starting their spring migrations up to three weeks earlier than usual and seeking summer breeding areas much farther north than in the past. Habitat changes are triggering earlier flowering and leaf-out times for plants, providing shelter and plant food for emerging insects and thus increasing the amount of insect prey for arriving flocks of birds in the early spring. For these avian ecological nomads, it's already a Greenhouse world!

How will we Boomers cope with that same Greenhouse world? Will we *move* to sensible retirement destinations, like snowbirds who are shifting their ranges? Will we *adapt* our lifestyles and strategies as well, ending up as ecological winners? Or will some of us be ecological *losers*, like the birds that may soon be going extinct? We have several advantages over them: We can understand the basics about global warming, we can research specific locations and lifestyles, and, above all, we can plan ahead. In part three, we suggest specific ways of planning ahead—strategies for living well in the age of Global Warming.

Part III
FINDING YOUR SOLUTION

"Another decade or so, and it'll be warm enough for us."

FINDING YOUR SOLUTION

t's time for a wake-up call. The Greenhouse world is real, it's right here, and it's right now. The carbon dioxide and other Greenhouse gases already in the atmosphere will continue to warm global climate to temperature levels not experienced for the last 100 million years. This accelerated warming will continue through at least the twenty-first century *no matter what* environmental programs are started today. Key times of intensifying Greenhouse environmental changes will coincide with each of the three Boomer Breakpoints, the demographic times at which Baby Boomers must make life-changing decisions about lifestyle and retirement destinations. The first Boomer Breakpoint in the year 2010 marks the time when the vanguard will begin to seek new lives in early retirement. The second Boomer Breakpoint in 2025 will be the peak retirement time for most of the 78 million American Baby Boomers, both native-born and naturalized. By the year 2070, the third Boomer Breakpoint marks the Boomer legacy, as our inspirations and our assets are passed on to the next generations, the Gen-Xers and the dot-com generation of Millennium Kids.

Now is the time to plan your personal passage through these critical Boomer Breakpoints. In the next three chapters, we offer tangible suggestions to help guide you through these critical decisions.

Chapter 8 helps you to develop an Ecological Survival Kit for planning what you want. What are the critical questions you ponder? How will you capture the good life? What are your priorities, your sensitivities, dreams, and fears? What do you need to know? Where can you go to find your own answers? What's your timetable of *personal* boomer breakpoints for making ultimate decisions?

Chapter 9 spells out a broad variety of strategies for adapting to the opportunities available in a Greenhouse world. We explore how you can plan effectively to become a Greenhouse winner, and we caution about the kinds of decisions (or inaction) that stack the environmental odds against you. You don't have to choose wrong, facing the rest of your life as a Greenhouse loser or as an ecological refugee!

Chapter 10 is about the future and how we Boomers can creatively shape our legacy for the Greenhouse world. There are lots of things we can do to reduce our contribution to habitat destruction, resource depletion, and degraded air and water quality, through our resource-consumption and lifestyle choices. Using renewable energy sources, a green approach to building design and energy conservation, and supporting of organic agriculture are major ways we can contribute. We can use eco-savvy planning for developing the communities in which we will live. Our personal actions can promote conservation of natural areas and ensure preservation of both our rich biological and cultural heritage. Simple solutions in landscape management can mean the ecological difference for survival of rare and endangered species of native plants and animals, as well as their fragmented habitats.

Knowledge is power. We Boomers can make a personal difference in shaping our own future, and in so doing we help shape the planet's future.

CHAPTER

8

Ecological Survival Kit

Aplan for action requires assembling a set of tools for coping with changes that lie ahead—an **Ecological Survival Kit**. With what you learn here, you can find out the right information to evaluate both beneficial and deleterious Greenhouse-world changes that will affect your chosen home location(s). As an *ecological stakeholder*, whether a homeowner, a concerned citizen, a real estate developer, or a city planner anticipating the onslaught of arriving Boomers, you need the right tools in order to cope with Greenhouse climate change and its environmental ramifications. We focus here on risk assessment, financial security, social security, and ecological security.

Packing Your Ecological Survival Kit

Your Ecological Survival Kit should include several basic tools. The first of these is a personalized list that includes the essential elements of the lifestyle you choose to live. This *priority list* includes what matters most to you: the optimal temperature range within your personal comfort zone; surroundings in which you feel most comfortable, including both natural landscapes and cultural amenities; lifestyle activities you wish to enjoy in retirement years; and the resource-consumption choices that reflect your values. Consider the level and kinds of risk you are willing to take. Think about the amount of resources you wish to allocate for ecological security, such as special FEMA opportunities for catastrophic flood insurance available for residential sites in river floodplains and coastal zones. These are fundamental decisions that will constrain to a great extent where you will be able not only to live, but live well.

The next tool in your Ecological Survival Kit is *knowledge*. You need access to a whole library of information for discovering facts and figures about the present-day and future environment for your chosen lifestyle and destination. Assembling this library is made less difficult today because much of the truly critical information can be found on the World Wide Web. Even if you don't have a personal computer with an Internet connection, you can access this vital information resource by computer through any public library. Your personal library should include a comprehensive list of Web sites, detailing the latest assessments of Greenhouse-world impacts resulting from climate change. Start with the list of electronic bookmarks kept updated at our Web site, http://boomerbreakpoints.com/.

You need the most current regional assessments, with Greenhouse-world maps showing the kinds of weather patterns and environmental risks you'll encounter. You need to know what you, your local community, and your government have available as successful adaptive solutions to living well in the Greenhouse world. Simply put, *you need answers*, and these are available. Key links to government agencies provide a wealth of landscape-level information on the weather, water and air quality, biota, and land use envisioned for the Greenhouse world.

Once your Ecological Survival Kit is packed, you will be ready to develop a plan for action—an **ecological survival strategy**—based on up-to-date information, best-reasoned contingency plans, and enlightened self-interest.

Contingencies

Contingency planning is essential for preserving hard-earned wealth, particularly in the form of real estate within areas prone to be hit by Greenhouse-world catastrophes. Economic ecologist C. S. Holling and colleagues have developed a novel approach to visualizing the world as an integrated system in order to evaluate **contingencies**, the suite of possible interactions among our complicated economic and ecological systems.[1] Formally termed **cross-impacts analysis**, this method is a quick way to identify the most important influences of climate change on your lifestyle.

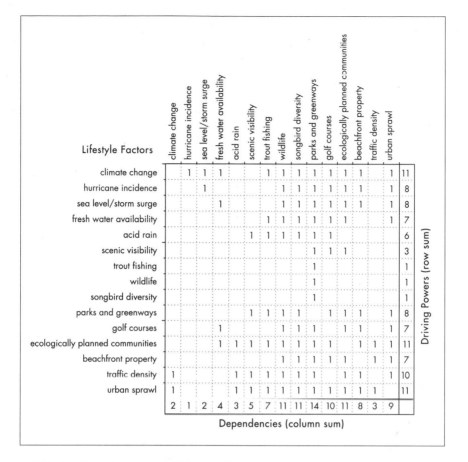

Table 8-1. **Greenhouse-world lifestyle chart**

Lifestyle Factors	climate change	hurricane incidence	sea level/storm surge	fresh water availability	acid rain	scenic visibility	trout fishing	wildlife	songbird diversity	parks and greenways	golf courses	ecologically planned communities	beachfront property	traffic density	urban sprawl	Driving Powers (row sum)
climate change		1	1	1			1	1	1	1	1	1	1		1	11
hurricane incidence		1						1	1	1	1	1	1		1	8
sea level/storm surge			1					1	1	1	1	1	1		1	8
fresh water availability							1	1	1	1	1	1			1	7
acid rain						1	1	1	1	1	1					6
scenic visibility										1	1	1				3
trout fishing										1						1
wildlife										1						1
songbird diversity										1						1
parks and greenways						1	1	1	1		1	1	1		1	8
golf courses			1					1	1	1		1	1		1	7
ecologically planned communities			1	1	1	1	1	1	1	1	1		1	1	1	11
beachfront property								1	1	1	1	1		1	1	7
traffic density	1					1	1	1	1	1	1	1	1		1	10
urban sprawl	1				1	1	1	1	1	1	1	1	1	1		11
Dependencies (column sum)	2	1	2	4	3	5	7	11	11	14	10	11	8	3	9	

Table 8-1 provides one version of such an analysis, personalized in the form of a **Greenhouse-world lifestyle chart.** This chart includes climate change as well as a number of other factors that we think are crucial to consider. These lifestyle factors affect in some way the kinds of environments and lifestyle choices we and other Baby Boomers look forward to enjoying. The same climate and **lifestyle factors** are listed both across the columns and down the rows of the chart, with a mark placed in each box of the *interaction matrix* where one factor potentially influences another. Certain factors have global scope; others are very localized.

For example, either directly or indirectly, ongoing climate change influences virtually every other aspect of the Greenhouse world. In contrast, another lifestyle factor, building and maintaining leisure golf courses, will probably impact many fewer aspects of cultural and biological communities. Watering needed to grow turf may require a municipal permit for another well drilled to tap into the potentially finite resource of a groundwater reservoir. The golf course network of mowed fairways, sand and water traps, and intervening forested roughs will determine the total number of species as well as the mix of native and exotic wildlife and even pests such as encephalitis-carrying mosquitoes seeking suitable habitat. Flanking the eighteen or so fairways, the spatial configuration of homes, driveways, and supporting roadways shapes patterns of traffic flow within and beyond a golf-centered development of leisure homes. By following basic rules of ecological landscape planning, environmental benefits and human needs can be balanced to craft sustainable, ecologically designed communities that maximize quality of life and minimize disease and depletion of natural resources.

For each row and column in table 8-1, you can easily tally the number of times that a given lifestyle factor influences all the other ingredients of the good life you value highly. You can then plot the position of each lifestyle factor on figure 8-1 to display your priorities for a **Greenhouse-world lifestyle**. Some lifestyle factors are so critical and regionally pervasive that they drive the ecological and cultural systems in which we live. These **driving factors** cause reactions or trigger consequences for other factors that are vulnerable to change, what we call the **dependent factors**. The Greenhouse lifestyle chart helps you distinguish between the driving force producing the action and the subsequent dependent reaction—distinguish, that is, between cause and effect. This chart plots each factor's value as a driving power (the total score across its row) against its value as a dependent factor (the total score down its column). The chart quickly sorts out for you which kinds of driving powers are beyond your control, as well as those driving powers you can shape by constructive action. It lets you know what you can do to enhance your local quality of life even in the face of regional and global change.

Here's how the factors sort out—as *forcing, relay,* or *result variables,* depending on which of three areas they occupy in the chart. The ultimate *forcing variables* tally up on the chart as having a combination of strong dri-

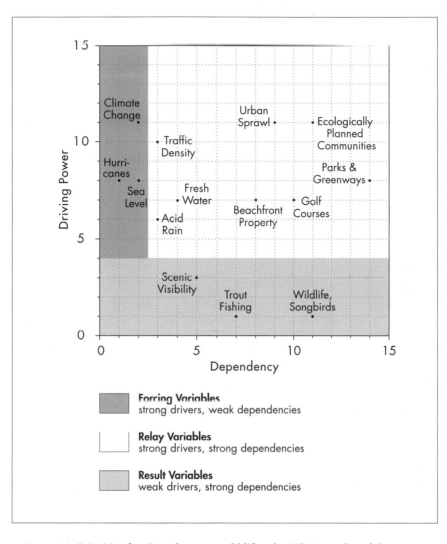

Figure 8-1. **Priorities for Greenhouse-world lifestyles.** This is a plot of the sums across rows (driving power) and down columns (dependent factors) for lifestyle factors listed in table 8-1. The higher the driving power of any factor, the greater the number of ways it has to influence your chosen lifestyle. Factors with very high driving power, such as global warming, may seem all-pervasive: they affect virtually every aspect of your lifestyle.

ving value and only a small dependent value (figure 8-1); these predominant factors shape overall quality of life and drive change within the Greenhouse world. *Relay variables* plot in figure 8-1 as having a combination of strong driving value and strong dependent value; these factors represent powerful feedback mechanisms that link humans to their living environment. The "relay" connection of these factors strongly links how we impact the environment with how our environment impacts us. The *result variables*, which tally as a combination of weak driving value and strong dependent value, are the ecological products, or results, created by Greenhouse-world changes together with cultural actions.

Strong driving powers (forcing variables in figure 8-1) include overall climate change, increasing traffic congestion, and exponential growth in total population density (not charted), as well as prevalent kinds of natural disturbance such as hurricanes, infrastructure capabilities for supplying necessary freshwater and electrical energy, and environmental pollution such as acid rain. These sorts of driving powers affect the course of urban sprawl and sustainable community development, the real estate value of beachfront and lakeshore properties, and the remaining biological diversity of wildlife and wildflowers.

It is apparent from table 8-1 and figure 8-1 that some of the most important and all-pervasive driving factors or variables that affect us, like global climate warming itself, are beyond our individual ability to control. We have to plan and adapt to them. However, we may be able to do something about the more immediate and personally relevant kinds of factors (the relay and result variables in figure 8-1). For example, if we are concerned about acid rain and the acidification of trout-fishing streams, we can join concerned citizens' groups to influence the political regulatory control of emissions from fossil-fueled power plants, reduce acid rain at its pollution source, or, locally, we may be able to help restore the natural pH of trout streams through adding chemical buffers to reduce the impact of acid rain.

For most of us, the most relevant factors for planning (the forcing variables in figure 8-1) are those that will affect our day-to-day lives, for example, climate change. Knowing this, we can get more specific—understanding how changes in storm tracks associated with accentuated El Niño/La Niña events, for example, will affect the location and duration of winter snowpack. This knowledge in turn may influence our weekend plans for cross-country skiing and snowmobiling trips or help us to predict how many dry sunny days will be available for enjoying the golf course or for backcountry treks.

You can modify the Greenhouse lifestyle chart to take into consideration the lifestyle factors most important to you. Based on information presented in chapters 4 through 7 and gleaned from environmental-change Web sites, you can work out the cross-linkages and see how they stack up in relative importance. Making your own Greenhouse lifestyle chart will help you to prioritize your goals and analyze your risks.

Strictly need to know

Knowledge can set you free—to make your own life-altering decisions. Timely, accurate, and relevant information is absolutely critical in your quest to understand, then to implement, your informed course of action.

Where can you go to find answers to your very specific and personal questions? For local, regional, and national perspectives, you need electronic access through several Internet portals.

Straight talk on global warming. For commonsense answers, start at the logical first stop on your whirlwind Internet tour—visit the virtual Visitor Center for Climate Change. You want the *Climate Change Outreach Kit* prepared by the Environmental Protection Agency.[2] Order their free CD for details and view online the general resources and advanced fact sheets. Using these online tools, individuals, community groups, and even state agencies can discover viable strategies and solutions, effective policies, and green technologies. As this electronic publication urges: get the facts, save money, improve air quality, and lower the environmental degradation that's risking your health. The more you learn, the more questions you'll ask.

Search by topic. For particular Greenhouse-world topics, the most comprehensive guide we know is compiled and updated by NASA. Their *Global Change Master Directory* (GCMD) can be reached online at http://gcmd.gsfc.nasa.gov/. This comprehensive resource will help you sort through the wealth of international studies and scientific databases to find the hard facts (or at least the most well-informed estimates currently available).

One of the best Internet clearinghouses for present and Greenhouse-world climate information can be found at the Climate Data Library (Web site at http:// ingrid.Idgo.columbia.columbia.edu/). The Global Environmental Facility of the United Nations (UNDP-GEF) provides an excellent, online *Climate Change Information Kit* as part of its program to help forge international cooperation and finance actions to address four critical threats to the global environment: climate change, biodiversity loss, stratospheric ozone depletion, and degradation of international waters.[3] State-of-the-art syntheses of vegetation response to Greenhouse climate change are provided by two comprehensive projects: the international scope of VEMAP, better known as the Vegetation/Ecosystem Modeling and Analysis Project (http://www.cgd.ucar.edu/vemap/ and http://www-eodis.ornl.gov/VEMAP/guides/users_guide.html);[4] and the Global Change and Climate History Project for the Western United States, coordinated by the United States Geological Survey.[5]

Search by information source. Check out the *Gateway to Global Change Data and Information System* (GCDIS) at http://www.globalchange.gov/. This gateway is coordinated as a public service outreach by the U.S. Global Change Research Program (USGCRP). Here, global change resources are offered as links to several kinds of information options and to thematic programs of all major federal agencies. Many of these agencies provide user-friendly syntheses of their taxpayer-supported research, a library of online publications and climate change databases, and an education section with suggested classroom tools and activities.

Search by geographic location. First, search by the state in which you live, or want to live. The *Global Warming Site* of the Environmental Protection Agency (http://www.epa.gov/globalwarming/impacts/stateimp/index.html/) is truly a national treasure. You can view the electronic file for your state in order to learn about local climate patterns and future changes. With thoughtful and informed reports expressed in commonsense language, Greenhouse-world impacts are spelled out as they will affect human health and quality of life, water resources, agriculture, forests, and wildlife. On the broader regional and national scales, important environmental summaries for the northern latitudes of Canada are available from its government site *Global Climate Change* (http://climatechange.gc.ca/english/html/index.html/).

For Greenhouse-world assessments and remedial actions at the international level, the ultimate voice for an integration of science, industry, and governmental policy is provided by the Intergovernment Panel on Climate Change, the IPCC (http://www.ipcc.ch/about/about.htm/).

For an extraordinary collection of environmental data sets and relevant maps, check out the continually updated array of CDs available for purchase by credit card from the National Geophysical Data Center (NGDC) at the National Oceanic and Atmospheric Administration (NOAA). This online store (at http://www.ngdc.noaa.gov/store) offers global digital imagery for land and ocean (CDs for Terrain Base and for Global Relief) as well as detailed coastal bathymetry for U.S. coastlines (CDs for the East Coast, and NOS Hydrographic Survey Data) and Great Lakes (CDs for Lake Michigan, Lake Erie, and Lake St. Clair). Most important is the two-CD collection Global Ecosystems, presenting an impressive suite of spatially detailed maps for Greenhouse scenarios of global climate warming and vegetation response. Go to http://www.ngdc.noaa.gov/mgg/fliers/se-2006.shtml/ to learn more about these Greenhouse environments as portrayed in joint eco-futurist projections by EPA and NOAA.

Search by the kind of imminent ecological danger. The PEW Center of Global Climate Change presents a wide-ranging suite of authoritative position papers. These excellent syntheses include state-of-the-science concerning climate change, as well as provocative reports on the kinds of Greenhouse-world dangers that threaten coastlines, agriculture, and freshwater supplies (http://www.pewclimate.org/projects/).

Do you live in a floodplain or on the coast? Are you at risk of being flooded out? Knowing the precise elevation of your home's foundation is critical. To get digital or paper copy of detailed (7.5-minute) topographic maps showing the precise elevation of your house site, check online for the maps published by the United States Geological Survey (USGS) at http://usgs.gov/mac/findmaps.html/. As an ongoing and collaborative effort between USGS and Microsoft, the Encarta Terraserver, located on the Web at http://terraserver.microsoft.com/place.asp/, provides an extraordinary map resource for the conterminous United States. You can download and print digitized versions of aerial photographs, topographic contour maps, and topographic relief maps, with spatial resolution zooming to as fine as one meter. In addition, SPIN-2 satellite imagery covers much of the United States as well as selected regions of Europe and the Middle East.

As an alternative to obtaining free topographic maps showing site elevations, you can go to the TopoZone Web site at http://www.topozone.com/. You can point and click with your mouse to pinpoint specific locations and read out their exact map coordinates in latitude and longitude and in UTM. To locate place names and prominent landscape features rapidly, use the USGS Geographic Names Information Service at http://www-nmd.usgs.gov/www/gnis/gnisform.html/—this site helps you find county names, populations, and elevations.

Will you be flooded out permanently by a Greenhouse-world rise in sea level? Find your own answer—calculate the near-term chart for rising sea level for your own coastal community at the EPA Web site http://www.epa.gov./globalwarming/publications/impacts/sealevel/probability/chapt_9.pdf/ (see chapter 4).

Learn about the potential flood hazard *before* you buy that ultimate dream home near the coast or along a river floodplain. The Federal Emergency Management Agency has already evaluated the present flood risk for *your home*. Surprised? FEMA needs to know its downside risk before agreeing to sell you catastrophic flood insurance. Your worry level goes up dramatically if your home is presently mapped in the Special Flood Hazard Area. These locations of heightened ecological risk currently have a 1 percent chance of being flooded in a given year. In the Greenhouse world, flooding risk is a reality and not some distant statistical probability. Even if you don't care, the bank holding your mortgage loan certainly does. The solution? At your county courthouse, inspect FEMA's maps (called Flood Insurance Rate Maps [FIRM] or Flood Hazard Boundary Maps [FHBM]). Too much work? Online, order your own set of relevant maps from FEMA's National Flood Insurance Program (http://www.fema.gov/nfip/).

If you live in the continental interior, perhaps the threat of tornadoes captures your attention. For climatological summaries of tornado data by state, check the Extreme Weather Sourcebook online at http://www.dir.ucar.edu/esig/HP_roger/sourcebook/report.html for geographic ranking by typical damage costs and by storm frequency.[6] For example, Greenhouse-world projections of the shifting locations of Tornado Alley show dramatic changes across the sunbelt (see the Tornado Alley Web site at http://www.ghcc.msfc.nasa.gov/regional/assessment_national.html).[7]

Use the Web to create a **custom hazard map** identifying the kinds of disasters you face. Through a National Partnership of FEMA and the Environmental Systems Research Institute (ESRI), you can identify individual communities and quickly obtain information about the potential for flooding, hurricanes, tornadoes, hailstorms, windstorms, and earthquakes. This information is found at http://www.esri.com/hazards/.

Search the National Assessment. Yes, you too can tap into the awesome power of the New Greenhouse Synthesis. This latest word, published online in the year 2000, scopes out the big picture for the conterminous United States, focusing on near-term consequences of global climatic, environmental, and socioeconomic change. The summary reports are presented in the Overview Document and in the more detailed Foundation Document (http://www.gcrio.org/NationalAssessment/foundation.html/).

This newest national assessment has been developed within the U.S. Global Change Research Program (USGCRP; see http://www.usgcrp.gov/ and http://www.gcdis.usgcrp.gov/pubs.nap.html). Substantive reports are being prepared for twenty key geographic regions (for the status of these reports, check http://www.nacc.usgcrp.gov/regions/): Alaska, Appalachians, California, eastern Midwest, Great Lakes, Great Plains (central, northern, southern), Gulf Coast, Hawaii and Pacific Islands, metropolitan East Coast, Middle Atlantic, Native Peoples/Native Homelands, New England, Pacific Northwest, Rocky Mountains and Great Basin, South Atlantic Coast and Caribbean, Southeast, and Southwest (Colorado River Basin, Rio Grande River Basin). Each of these regional syntheses provides Greenhouse-world projections for the next twenty-five to thirty years, covering the Boomer Breakpoint in 2025, and for the year 2100.

The national-level reviews in the Overview and Foundation Documents emphasize key issues related to human health, agriculture, forests, coastal areas and marine resources, and the environmental quality of air and freshwater resources. The regional assessments concentrate on the needs of stakeholders, establishing constructive dialogue among the concerned public, scientific communities from universities and conservation groups, private industries, and government from all levels.

Based on current scientific knowledge, each region-specific assessment translates this understanding of climate-related issues into programs for action and adaptation. The regional summaries explore both vulnerabilities and opportunities for response to the enhanced variability of changing climate and environments. For example, the Assessment Report for the Southeastern United States is being coordinated through the Global Hydrology and Climate Center (GHCC) at the University of Alabama at Huntsville and through Florida State University. An extensive "white paper" available online explores the critical issues, concerns, and stresses shared by regional stakeholders.[8] Such geographic syntheses pose the leading questions about Greenhouse-world impacts, the status of available scientific knowledge and data sets, the necessary directions for future research, and viable solutions for coping with, adapting to, and exploiting climate change. The Southeastern Assessment explicitly considers Greenhouse-world responses for eight kinds of potential impact, including extreme climate events (such as El Niño/La Niña) that influence hurricane and tornado activity, urban areas, and human health, with a special emphasis on water resources, air quality, agriculture and forestry, coastal resources and freshwater fisheries, and parks and public lands.

Ask Dr. Global Change. Of course, as your last-gasp attempt to find answers, you can always just "ask Dr. Global Change." Advice from Dr. Global Change can be found at http://www.gcdis.usgcrp.gov/help/ask-doctor-form. html.

These online resources provide what you need to know *just in time* for you to make hard decisions about your realistic lifestyle options.

Hard Decisions

Making decisions is all about risks, rewards, and time frame. Balance the risks you're willing to take against the rewards you strive to gain and the time window available for action. Your tolerance as a risk taker is conditioned by what you have at stake to lose. As a stakeholder, your assets may become threatened, even destroyed, by these risks, which vary in type, immediacy, and magnitude.

This problem quickly becomes a matter of your personal tolerance to risk: high, intermediate, or low tolerance, as well as avoidance—which means you have a low mental-pain threshold and thus low tolerance for even the mere prospect of loss. Some individuals have aggressive personalities. They deliberately seek risky ventures and the adrenaline rush of gambling everything they own to win big. These "opportunists" thrive on high risk in order to capture the highest prize in life's sweepstakes. Their mantra is "no pain, no gain." Almost inevitably, they try and fail. Usually, they go on to try again. Such people are resilient and appear to be emotionally strengthened by surviving intense "bottleneck" crises. They are thrill seekers, for whom each disaster and collapse of personal empire becomes a springboard for bouncing back and rebuilding, bigger and better. These **Ecological Thrill Seekers** should excel in the high-risk/high-quality opportunities of a dynamic Greenhouse world.

Others of us accept only more limited exposure to the risks we're willing to take. Many of us limit the downside potential for asset loss but also the upside potential for gain. We select safer opportunities for which we'll invest our time, energy, and fiscal assets. While preferring to stack our odds toward success, we feel that we can tolerate only intermediate levels of possible loss in order to enhance our likelihood of moderate gain. We call these Boomers the **Comfort Seekers.** They subscribe to the "Goldilocks" model of decision making in their quest for opportunities that are "not too hot, not too cold, but just right."

Still other Boomers have a much more conservative bent. Their philosophy of low risk, low pain tolerance is linked to the knowledge that time is precious. These Boomers, who are **Stability Seekers,** deliberately try to reduce their exposure to financial, social, or ecological catastrophes. They are willing only to take small risks, directly in proportion to their perception of a diminishing lifetime left in which to recover from losses. Stability Seekers fend off the world of tempting thrills with an attitude of "It may not be much, but it's all mine, owned free and clear of debt!" Wary of even the exuberant (if not frothy) stock market, Stability Seekers are entrenched stakeholders, not speculative stockholders. Their defensive mind-set is quite straightforward, one that is focused on protecting their limited claim on the good life. Stability Seekers hope to find one **safe site** or refuge, one haven, that will be buffered from ecological change in an uncertain and potentially hostile Greenhouse world.

Risk: Changing Perceptions or Changing Realities?

Whether you're a Thrill Seeker, a Comfort Seeker, or a Stability Seeker, the ecological risks you understand and try to plan for may be very different from the new Greenhouse-world reality. The discrepancy between your *perception* of risk and the near-future *realities* of risk may make you ecologically vulnerable. You need to cultivate flexibility for unexpected kinds of natural disturbances and ecological surprises. *It's what you haven't yet thought about that may become most important* in ruining your financial investments, disrupting your social milieu, and degrading your ecological quality of life. Risk factors and their many complex interactions are dramatically changing in what is now the fast-evolving Greenhouse world.

In academic ecological circles, we talk about an **intermediate disturbance hypothesis.**[9] Biological opportunities increase dramatically for species that can adapt successfully to repeated, intermediate-scale disruptions in their environment. Intermediate-scale means that the disturbance, whether it be windstorm, wildfire, or landslide, occurs at only a moderate magnitude and with fairly long times between recurrences. Plant and animal populations are thus exposed to physical harm only occasionally and are not threatened by complete extinction. Hence they can recover rapidly from minor setbacks.

In the Greenhouse world, what kinds of natural disturbances can Boomers expect to be exposed to and still recover rapidly? How can we rate the relative influence on our future lives of events such as floods, droughts, hurricanes and tornadoes, blizzards, glaze-ice storms, sustained rainfalls, wildfires, volcanic eruptions, and even earthquakes?

Today, each geographic region has a distinctive climate that directly shapes the kinds, magnitudes, and recurrence intervals between such disruptive events.[10] In choosing your lifestyle and retirement destination, you are also selecting its related package of potential disasters, whether you are aware of it or not. As climate regions shift geographically over the next several decades, there will be changes in the disaster regime at any one location. For example, wildfires may become more frequent and more severe across the coastal pinelands of the southeastern United States.[11] There, regional drought leading to forest dieback will cause a buildup of fuel load, dead branches and trees, just waiting for a lightning strike or a careless camper's match. Many environmental changes like this in the Greenhouse world will shuffle the ecological cards, making for an ever changing kaleidoscope of potential risks that Boomers will face in the years ahead.

Your perception of risk may track the demographics of the aging Baby Boom generation. The silvering of Americans, added perhaps to their greater base of lifetime experience as they age, may correspond with more conservative tendencies toward risk taking. As such, our thresholds of risk avoidance may telescope toward a personal need to exert greater control over our immediate situation.[12] As the new century unfolds, we may find ourselves progressing through stages of decreasing tolerance to environmental risk, shifting from ecological thrill seekers to comfort seekers and eventually becoming stability seekers.

Financial Security

John Bogle, in his widely praised book, *Common Sense on Mutual Funds: New Imperatives for the Intelligent Investor*, endorses a disciplined approach toward long-term acquisition of personal wealth.[13] He suggests *not* relying on only one **nest egg**, that is, one type of financial asset such as bonds, stock equities, or real estate investments. Rather, he champions a style of **asset allocation** that diversifies within a collection of nest eggs, spreading the potential risk of asset loss across a whole suite of financial investments. To minimize financial risk in stocks, for example, Bogle encourages investing in the kind of indexed mutual funds that he first created in 1975 at the Vanguard Group of Investment Funds. In this plan, individual investors pool cash and risk with many other investors, together buying shares of stock from many kinds of companies reflecting diverse product lines and services. The ups and downs of this kind of mutual fund track a representative mix of companies, providing a convenient "index" of the overall status of an industry, a national economy, or even an investment philosophy. For example, various index funds may limit themselves to technology stocks (one industry) or the stock market as a whole (one nation) or stocks with accelerating profit growth (one investment philosophy).

Bogle carries this line of logic further, in recommending regular monthly investment in a variety of nest eggs, each egg accumulating within one of several **baskets**. One basket may hold your bond funds and another your REITs (Real Estate Investment Trusts). Yet another basket contains your North American stock equities. This domestic basket may contain funds that focus on the narrow Dow-Jones index of 30 industrial firms or the broader S&P 500, an index for "blue chip" stocks of the 500 largest companies. Alternatively, you may select an index fund based on the Wilshire 5000, which tracks virtually all U.S. companies with stocks that are regularly traded. Geographic coverage can be expanded as well to reduce your vulnerability to financial risks in any one area. Just add more global baskets for companies in other developed countries, and perhaps even a basket for index funds investing in more speculative areas called emerging markets. And how about adding to this list a basket portfolio of "socially responsible" investments—screened, that is, to *exclude* companies whose practices or products you find objectionable, or specifically to *include* investment in companies whose socially or environmentally progressive policies, products, or services you wish to support?[14]

According to John Bogle, your ultimate financial success is determined largely by your initial asset allocation. This simply means how you decide to distribute nest eggs among the many kinds of baskets you've chosen. The beauty of using this approach is that different kinds of assets, industries, and national economies move independently. The wise investor exploits this fluctuating nature of free markets precisely because of their lack of synchronism. If one market basket should free-fall and crash, not all of your nest eggs are scrambled. **Financial security** is simply not putting all your nest eggs in one basket.

Social Security

For many Boomers, our sense of well-being and **social security** (meant in a broader sense than just the federal government's pension plan), our major capital outlay in monthly expenses, and our primary strategic resource intended to fund retirement are all tied to one key asset, the family home. As discussed in chapter 2, the McFadden doomsday scenario (figure 2-3) predicts a collapse of the national housing market starting after the Boomer Breakpoint of the year 2010. The combination of plummeting prices and probable difficulty in selling your home may create a personal crisis of illiquidity. Your hopes for the one nest egg of your home may be badly cracked if not broken and scrambled.

Obviously, if this scenario is accurate, you have a personal incentive to "cash it all in" *before* the rest of the Boomers retire (figure 2-4). After 2010, retiring Boomers will competitively bid up real estate values in desirable retirement destinations, while values continue declining in other areas. But you may not be able to act just yet. Your quest for social security may be limited by the number and size of nest eggs you've accumulated. Not enough cash may mean you can't move! Yet you can still plan ahead.

Let's look on the bright side, however far ahead it may be. You're ready and able to move on to a new lifestyle in a new place. The questions are intriguing. Should you put all your nest eggs in one basket, one home—that is, the All-American Dream House with all the comforts and energy-smart amenities? Perhaps you would prefer an even swap, trading your urban home, condo, or whatever you've got for land, for more elbowroom in the countryside. Do you seek a *sense of place* on a farm or ranch? For those who champion a green lifestyle characterized by voluntary simplicity, this may mean a new start, converting to nonelectric self-sufficiency on a forty-acre woodlot or a renovated historic farmstead. Perhaps you seek the kind of "conservation ethic" lifestyle described in Aldo Leopold's compelling *A Sand County Almanac.*[15]

Alternatively, should you choose a few baskets, each with an equal-size egg of investment? Will two (or three) seasonal homes of intermediate cost be a better option than one expensive home or rural homestead?

Perhaps, for you, the spice of life comes from a variety of brief but intense experiences—a predictable seasonal round of leisure activities, shopping extravaganzas, and gourmet restaurants. A third possibility is to acquire many small baskets, each with a small egg. You may wish to spread your risks among small homes in many locations—if one cottage is damaged or becomes undesirable, there are always plenty more bungalows. After all, that's why you pay for home owners' insurance!

A fourth option is to buy no basket at all. Rent where you wish to live, or be a moving target in a Recreational Vehicle (RV) rolling down the road. Let other people risk their capital in your rented home, paying taxes and insurance.

Whatever you decide, the demographic pressure of relocating Boomers may exacerbate infrastructure problems in booming resorts and quaint "rediscovered" communities. So higher taxes will surely follow. As economic refugees flee the Midwestern Rustbelt and ecological refugees escape the Southern High Plains, the property tax adjusters are preparing new styles of creative accounting. Seasonal migrants may try to minimize taxes by choosing to keep legal residence at another home place. Yet local assessors have started to erect a property tax system with three tiers, as we have already experienced with our summer cabin in Upper Michigan. Year-round locals with their Homestead Waivers are granted lowest property tax rates. Weekenders coming from other parts of the state of Michigan are paying a value-added tourist tax. Outlanders, whose primary home is in another state, have no local vote or representation; they are taxed most liberally. Perhaps at least grudgingly, we should acknowledge that there *may* be a defensible logic behind such overtly exploitationist schemes.

In all fairness, the vested interests of established, year-round residents in any community must be protected. Perhaps an "impact cost" should be levied to help share the burden of Boomer resettlement, as summer resorters and new arrivals tap into municipal utilities for drinking water, sewage, electricity, and telephone/cable/Internet connections. Desert sunbelt communities require a fee of as much as $50,000 for impact costs *before building permits are issued!* In Michigan's Upper Peninsula, local communities actually *enhance* their quality of life from the sumptuous tax subsidies contributed by seasonal, nomadic snowbirds.

These local tax boards will face a Greenhouse-world quandary. Should property tax assessments be used to discourage all but the most affluent Boomers, thus placing subtle limits on future growth? Can the broadening tax base be used to plan for and construct more infrastructure capacity, buffering against the ecological vulnerability of water and energy reserves? Can the three-tiered tax system offer creative relief for long-time local residents, such as protecting retirees from booming real estate valuations that would otherwise inflate tax bills and force them to sell out? Or, ever pragmatic, should the tax boards continue business as usual, simply exploiting the fabled goose as relocating Boomers build their ultimate dream homes and lay their golden nest eggs in the local economy?

Ecological Security

Where you live determines how well you can live. Your choice of home place(s) shapes your immediate **ecological security** as well as your future vulnerability to ecological risk. Your flexibility in Greenhouse-world options is dictated by the number of homes you own, their market value, and their geographic location (unless you're an RV nomad). You maximize flexibility at any one site by paying attention to many features, such as how a house is situated for passive solar benefit, or insulated to conserve on heating/cooling energy use, or elevated for flood protection. Have you planned well, or guessed wrong, or will you soon wake up to Future Shock?

Where you live (or plan to live soon) means that you've made your decision, or best bet, on the future. Is your game plan still open to a few good ideas? Chapter 9 gives our suggestions to help you refine your strategy for living well in the age of Global Warming.

CHAPTER

Ten Best Strategies for Living Well in the Age of Global Warming

You've packed your Ecological Survival Kit. You've worked up your chart of options for Greenhouse-world lifestyles. You know what you want to do. You've also decided on the amount and kinds of risk you're willing to take. Now the fun begins! Let's craft a Greenhouse game plan, combining your intended Boomer lifestyle with your priority list of attractions drawing you to a specific location.

Priorities

What comes first on your list? In the August 1999 issue of *Smart Money,* reporter Clifton Leaf checked off the expected big items people care about most in their retirement years: people, health and leisure time, affordability, and place.[1]

Our enhanced Boomer longevity means *more* **people.** More of us will stay around appreciably longer. The U.S. Census Bureau projects the "Floridization" of America by the year 2023, the date by which the rest of the nation will achieve the same proportion of senior citizens as live in Florida today—19 percent of the population. By the year 2030, the number of senior Americans age sixty-five or older should double to 70 million.[2]

Being surrounded by lots of people in your own age group often triggers one of two opposing urges: "let's party" or "it's time to move on to another place." The first urge, or *fun syndrome,* means you may desire in retirement all the cultural diversity you've come to expect of a major metropolitan area. For some, the good life also means quick access. Don't miss a thing. Be right in the midst of all the action. No more long commutes driving into the sun or trapped in gridlock. One fashionable trend is to return to the city center, refurbishing, for example, Victorian mansions or townhouses to their former glory. Some new communities are emulating scenes of the past, clustering picket-fenced homes along tree-lined streets within easy walking distance of shops and businesses.

In contrast, the *flight syndrome* reflects a psychological need to flee population centers, to find solitude in the richness of nature's biological diversity. Retirees of age sixty and older are flocking to rural landscapes, populating open green spaces at twice the rate of urban areas.[3] While going back to the land, affluent Boomers are also seeking a wilderness experience, but not necessarily camping out in tents. The *civilized rustic* lifestyle is one in which you can view wildlife from your cabin window while still enjoying the benefits of running water and electricity, which allows access to a satellite dish and the Internet.

As for **health and leisure,** *how* we spend our time and energy will shape *who we become.* Clifton Leaf tells us that by the year 2030, Americans during their entire lifetimes will spend, on average, more than half their waking hours in leisure activities.[4] Whereas many Boomers are intent on early retirement at age 55 or 59½, the next wave of Gen-Xers have already made it clear that age 40 is a more desirable goal for retiring to live well. In the twenty-first century, we'll certainly make up fast for those early years spent in school and at the workplace.

Our choice of *where* to retire will also shape who we become. Do you plan to live in one place for all four seasons, or do you choose one season for all places? Will you track one comfort zone of pleasing temperatures and sunny days, or do you enjoy a place with four distinctly different seasons? Each sort of choice opens up very different recreational opportunities. "Leisure activities" do not, however, necessarily imply consuming more or even spending money. Volunteering, community service, artistic endeavors, gardening, tending to home maintenance—these are also leisure activities suitable for retirement years.

When Clifton Leaf pondered the **affordability** of what we want, and how much cash we will have, he saw a growing gulf between Boomers. An intragenerational chasm separates those who *have saved* and those who *have not yet saved* for life after retirement. One Wall Street analyst predicts that the Dow-Jones Index will soar to 41,000 by the year 2008. Unfortunately, the flip side of this prediction is that it will then collapse 70 percent in value by the year 2020. Private employers are restructuring their pension plans to limit their downside exposure as well. Cost-effective changes in corporation pension plans may jeopardize or seriously reduce the amount of retirement benefits that Boomer employees can expect.[5]

The Social Security Administration is required to project its fiduciary responsibilities seventy-five years into the future. In this "pay-as-you-go" system, Social Security trust funds will be exhausted just as the youngest Boomers turn sixty-six and expect full retirement benefits (figures 2-2 and 2-4).[6] According to one survey, Gen-Xers believe they have a greater chance of seeing an alien spaceship from outer space than of being able to collect retirement benefits from Social Security.

Richard Leone, president of the think tank known as Twentieth Century Fund, attempts to reassure Boomers and Gen-Xers. He states:

> The American people have been treated to a steady diet of alarming "news" about the state of their most basic [public] pension program, the Social Security Trust Fund. But, like the rhetoric about employment, most of the concern is overblown. Social Security is currently running a surplus, and even when that runs out in about thirty years, the program would still be able to continue at *75 percent of current benefits, indefinitely* [italics added]. Confounding the confusion is the persistence of perhaps an even greater misunderstanding about private pensions. Although the assets of these plans have grown enormously, now totaling over $5 trillion [in 1997], they still are available to less than half of American workers.

More specifically, Leone adds, about 60 percent of all American workers are *without access to any sort of pension plan* beyond Social Security. Leone concludes that "most American workers simply lack financial assets" to provide for a decent retirement, since "85 percent of such assets are in the hands of the richest 10 percent of the population."[7]

Nevertheless, Ken Dychtwald, author of the Boomer manifesto *Age Power*, considers the ticket to affordability could be having that well-paying job and pension plan with a major corporation.[8] This ticket may be stamped null and void, however, if you can even get it. Boomers typically change jobs every three to four years, often well before the five to ten years required until their pensions become vested, so that the employing corporations take back their pension promises. Should Boomers be so lucky as to have vested pensions, the great majority cash in their retirement stash early for a lottery ticket, or that new boat, a vacation, or a down payment on a home. They've already spent their stake in affording the future.

Clearly, Boomers' uncertainty about their financial future represents a prominent concern in implementing a successful game plan for retirement. When and how much cash you can tap from your retirement assets sets the cash-flow constraint of what you will be able to afford. And, of course, a final factor is how much money you actually need to achieve your chosen lifestyle. If you make certain choices, it doesn't have to be a huge amount. Voluntary simplicity is one viable, thrifty lifestyle option that increasing numbers of Boomers and other Americans are taking. Your personal priorities, and how you manage all these issues related to affordability, will determine whether the other lifestyle and retirement decisions you make are your choice, are an acceptable compromise with some limited choice, or are your default options taken with no choice.

Ultimately, it all comes down to a matter of **place.** But which place? How can you choose your next destination, given that your finances and game plan permit it? Beware the promotional hype. Several sunbelt states are actively competing for the attention of the *gray gold,* the next wave of affluent retirees seeking a new start.[9] Living well translates for each place into basic issues of economic momentum, educational and cultural opportunities, recreational diversity, and, to quote *Money* magazine, "safety—on the streets, in the air and in the drinking water."[10] Increasingly, our priorities for finding the best place to live must explicitly factor ecological risk into the personal equation for decision making.

Risk Takers in a Greenhouse World

In the Greenhouse world, the degree of climate change expected, the kinds of natural catastrophes we face, and their impact on social infrastructure (electrical power, water supply, transportation, leisure activities, etc.) altogether define our exposure to ecological risk. Our potential for achieving a high quality of life depends on how well we can match appropriate strategies to apparent risks. Table 9-1 shows a variety of lifestyle solutions that result in both winners and losers in this Greenhouse world. The strategy that's best for you depends largely on your tolerance for risk. Here we identify three Greenhouse-world categories of personal risk tolerance: the Ecological Thrill Seekers, the Comfort Seekers, and the Stability Seekers.

Ecological Thrill Seekers: Adventurers

As a group, **Ecological Thrill Seekers** opt for a high quality of life balanced precariously on the threshold of extreme exposure to ecological risk. This combination of *high quality and high risk* challenges, even stimulates, Boomer entrepreneurs. They seek out both the most risky geographic areas and the times of greatest environmental variability in order to maximize their adrenaline rush as well as potential profits. Their objective is to win big at the "Greenhouse game of life."

We see three very different and successful Boomer lifestyle strategies for these thrill seekers: the Ecological Contrarian, the Ecological Speculator, and the Cash Flow King. As deliberate lifestyle choices, these Greenhouse-world philosophies are far from simple. Such schemes for making money require you to take advantage of your own insight and initiative, along with other peoples' lack of foresight or their ecological vulnerability. There are three ways to win big at this game. They are, however, strategies that carry with them high ecological risk.

The Ecological Contrarian

The Ecological Contrarian *does not necessarily* choose to do the opposite of everyone else. Rather, the essence of the **Ecological Contrarian strategy** is to discover and act on possibilities *before* they are recognized by the rest of the crowd. Remember the basic investing rule of Boomernomics: "get there before the rest of the Baby Boomers."[11] For example, many Boomers want to move to a lakeshore destination they have chosen for retirement. Appreciably younger, impatient members of Generation X, however, don't want to wait that long! Property values for lakeside bungalows have already soared beyond reason for the tens of thousands of natural lakes scattered from Minnesota to Maine. For some Boomers, it's a case of saving too little and

EXPOSURE TO ECOLOGICAL RISK	GREENHOUSE WINNERS WITH A HIGH QUALITY OF LIFE	GREENHOUSE LOSERS WITH A LOW QUALITY OF LIFE
High Risk	Ecological Thrill Seekers: Adventurers • The Contrarian • The Speculator • The Cash Flow King	Ecological Refugees
Intermediate Risk	Comfort Seekers: Ecological Nomads • The Snowbird • The Perpetual Beach Walker • The Time-Share Timer • The Sun-Seeking Road Warrior • The Seafaring Vagabond	Ecological Survivors
Low Risk	Stability Seekers: Year Rounders • The Homesteader • The Sustainable Hedonist	Ecological Losers

Table 9-1. **Ten strategies for Greenhouse-world winners**

starting too late, only to be pushed aside by aggressive Gen-Xers who move in and set up shop instead.

Here's a second chance to place in that good life at the lakeside. One viable contrarian solution is to acquire cheap lake frontage in a place not yet discovered by the other 78 million American Boomers. Within the next ten to thirty years, dramatic climate warming will occur across the northern Great Lakes and New England as well as throughout central and southeastern Canada. Climate zones will shift some three hundred miles farther north, bringing balmy summers to central Ontario and southern Quebec. Just imagine the hearty Minnesotans from fabled Lake Wobegon trekking northward to the endless boreal summers of Ontario's Lake Nipigon.[12] But who will get there first? Gen-Xers will also be emigrating into southern Canada and using the World Wide Web to broker the next great landgrab electronically. These Greenhouse-world entrepreneurs will use the Internet to promote lucrative land deals, lease long-term or sell the real estate, finance the mortgages, and build the Boomers' dream homes. Ever optimistic in a contrary world, ecological contrarian Boomers are also preparing to pan for gold in the next great rush around 2010. They will happily take their share of the *gray* gold as other maturing Boomers belatedly discover beautiful lakes, great fishing, and cheap land just north of the border. It's a win-win solution for ecological contrarians as fellow Boomers migrate northward along with the shifting climate and actually *enjoy* living in the Greenhouse world.

The Ecological Speculator

The **Ecological Speculator** provides the Boomer crowd with what they want now. This strategy stands in contrast to the bold contrarian style of providing what the crowd hasn't thought of yet. Many Boomers want it all now: the nicest properties residing in the most desirable, prestigious, and conspicuous locations. After all, what's the use of having wealth if nobody else can see that you have it? The Ecological Speculator plays to these Boomers' egos, their passion to take advantage of a good deal, and, most important, their ample bank accounts, which must be tapped in an artful way. We think of the strategy for this kind of Ecological Thrill Seeker as *deliberately tracking disturbance*. In ecological jargon, we would call these individuals "opportunistic recolonizers." They are the first to re-establish, as pioneer settlers, on sites devastated by some natural disturbance.

The strategy of the Ecological Speculator is elegant in its simplicity: buy low, sell high. Opportunistic Boomers and Gen-Xers will come in after an environmental catastrophe and buy up land and demolished homesites at depressed, fire-sale prices. They'll rebuild to current construction code and offer these value-added homes at extravagant prices. And they'll expect the arriving hordes of Boomers to beg for the first option to buy.

What is the reasoning behind this compelling marketing plan? If they can afford desirable real estate, many Boomers want only the best possible location—for which they are prepared to pay premium prices. Such affluent Boomers want *that* seaside home at the edge of the sandy beach. At the very least, they want a panoramic ocean view from their rooftop patio or upper-level balcony.

Here's the selling pitch for the coastal resort developer, which we hereby disown in case it's illegal:

The ocean shoreline is a finite resource. You know, they don't make any more ocean frontage today. When this last beach home is sold, there'll be no more put on the market. The neighborhood is full of successful people like you, all here to stay for at *least* another forty years. Don't expect any turnover in these exclusive homes for many decades to come. Remember, this is your LAST CHANCE. Even at this price, the value's bound to go up. It's an *investment*, really, you'll be amortizing over the next thirty years—see, it's good value, not expensive at all! If you're not ready to sign now, please move aside, and don't block the line of buyers standing right behind you.

Of course, when the wary prospective buyer asks about recent hurricanes, storm surge and such, the Ecological Speculator's reply is already programmed:

What's that you ask? Hurricanes here? Not to worry. Just remember hurricanes, or was it lightning, or tornadoes—whatever. They never strike in the same spot twice. Forget about our recent unpleasantness. That was *the* hundred-year storm. We've rebuilt, and this home is safe for the next *hundred years!*

For this high-quality/high-risk strategy, the Ecological Speculator has two viable options. The first is to hopscotch from one coastal disaster zone to the next. FEMA subsidies will supplement your capital outlay for financing the costs of home "reconstruction." Alternatively, check out figures 4-2, 4-5, and 4-6. If you live along the coast in Texas or Florida, don't bother moving. Just wait a few years until the next "hurricane of the century" strikes and creates your next Greenhouse-world opportunity.

The Cash Flow King

Nothing to lose is the third option for the Ecological Thrill Seeker. The Cash Flow King follows the **SOS strategy**, where SOS stands for "save our savings." Live the good life in high-quality surroundings, but close the front door and leave when environmental disaster threatens. For this strategist, note the important distinction: High-quality/high *ecological* risk pairs with low financial risk of your personal stake. The key is to minimize what is personally at stake, to consider that Greenhouse-world disasters are truly somebody else's problem. The Cash Flow King will minimize capital outlay in their home or homes so that they can maximize protection of other financial investments. Paradoxically, this risky ecological setting for your choice to live is counterbalanced with a conservative financial position with limited vulnerability as "stakeholder."

This solution to Greenhouse-world climate change involves living in the most desirable locations by *renting* a home for the most enjoyable times of year. What is the reasoning behind this? It is an unfortunate fact that some Boomers were born later than others. These latecomers showed up on the baby scene between the years 1957 and 1964 and have been playing catch-up ever since. They too crave that seaside (lakeshore, mountainside, sunbelt) dream home for retirement, but they have to pay their dues working for quite a while to come (see figure 2-2). They see real estate values escalating at a higher rate than their nest eggs are accumulating. Many latecomers are making a preemptive strike so as not to be excluded. They are buying their dream homes now but financing this dream by renting them out. The rental income effectively pays the monthly mortgage bill.

Think of the Cash Flow King as giving a somewhat *altruistic form of self-help* to these latecomer Baby Boomers. By renting, the Cash Flow King is relieving the owner's cash flow problem. He or she is actually helping the Boomer latecomers to capture that elusive retirement home while the possibility is still affordable. The latecomers, however, take on all the financial risk—the initial capital outlay as well as ongoing commitment for mortgage payments, FEMA flood insurance, home maintenance, property taxes, and so on. Much more critically, the latecomers also take on all the prospective high ecological risk. Cash Flow Kings plan to live well but on somebody else's liability. This is as close as it gets to a free lunch!

These three strategies for Ecological Thrill Seekers use high exposure to ecological risk to ensure a high quality of life. They creatively use knowledge of how, when, and where Greenhouse-world climate and environments will change. These Greenhouse winners will employ this knowledge to anticipate, adapt, and respond, shaping the local consequences of global warming to their advantage. The flip side of high ecological risk is one default option for the Greenhouse losers (table 9-1). **Ecological Refugees** are unknowing high-riskers who don't realize when and where catastrophes are most likely to strike. Not having planned will leave them refugees, running too late with too few options.

Comfort Seekers: Ecological Nomads

For most of us, life is all about acceptable compromises. We seek *high quality of life with only intermediate ecological risk* (table 9-1). We don't care for extremes. For instance, Hazel prefers daytime temperatures between 65 and 80°F, while Paul prefers cooler temperatures, somewhere between 55 and 75°F. In an effort to maintain our happy marriage for the long term, we plan to track our mutual comfort zone, 65 to 75°F, by migrating seasonally with the snowbirds. We are **Comfort Seekers,** planning a retirement lifestyle as **Ecological Nomads.** Our personal solution is to live the nomadic life of snowbirds.

The Snowbird

As **Snowbirds,** we choose one season to enjoy at all our places. We don't ask for much—just our reasonable compromise of 70°F for daytime highs, sunny skies, balmy breezes, pleasant hiking trails, endless sand beaches to stroll, and a bottle of good white wine to sip on the verandah while we savor sunsets! Ah, but how will we do it? With nest eggs split up in three medium-sized baskets. With three homes of modest value, strategically located, we spread our ecological risk and move every few months to linger in the best (for us) time of year for our wilderness forays.

Our personal priorities start with creature comforts—the appropriate range in ambient temperature. Next, we choose proximity to extensive natural areas with biological preserves and parks that protect the biological diversity of native species. On our annual sojourn, we hope in retirement to experience the rich spectrum of subtropical maritime ecosystems in the southeastern United States, the spectacular temperate deciduous forests of the Appalachian Mountains, and the boreal evergreen forests, freshwater lakes, and bogs of the northern Great Lakes region. Using our own research on prehistoric times of global warming, we've learned lessons from the past to help prepare for our future. As eco-futurists we're studying projections of the Greenhouse world. And we're applying what we've learned looking backward and forward in time to the immediate problem of where we will choose to live in retirement. For us, one optimal schedule is to spend *four months* mostly of winter in the coastal South (middle November through middle March), *two months* of springtime in the southern Appalachians (middle March through middle May), *four months* of late spring and cool summer in Upper Michigan (through middle September), and round-robin back with *two months* of warm fall weather near the Great Smoky Mountains (middle September to middle November). This is our strategy for a nomadic Snowbird lifestyle, giving us four months at each of three homes through the year.

The Perpetual Beach Walker

Katherine Pearson, the editor of *Coastal Living*, shapes the whole format of the bimonthly magazine with articles about the East and West Coasts of the United States that "follow the seasons from North to South and back again over a year's time."[13] Following the annual migration path of whales, the essence of this strategy for the **Perpetual Beach Walker** is to travel seasonally between two coastal locales, flitting from summer place to winter haven. For some Boomers, this annual odyssey may alternate latitudinally between California and Washington State on the Pacific Coast. Others may travel north and south between homes along the Atlantic Seaboard. Yet another bicoastal possibility is one that some of our summer friends have adopted— they start from winter refuges in coastal California or along the Gulf of Mexico and rendezvous each spring at their second, summer homes on the shores of the freshwater seas, the Great Lakes.

Whether marine or fresh, large expanses of open water ameliorate temperature extremes, bringing cooling sea breezes and an equable clime for those who seek comfort in a seaside lifestyle. Already, half of all Americans live in counties that front along the twelve thousand miles of U.S. coastline. This bicoastal strategy builds on a positive situation for many Boomers— where you live today can set the stage for creatively adapting this Greenhouse-world solution.

Even in the Greenhouse world, enhanced climatic variability will be damped by the buffering presence of great bodies of water. Coastal zones will continue to provide critical shelter for communities of Boomers as well as otherwise-threatened species of native plants and animals. By the Boomer Breakpoint 2010, more than 70 percent of all Americans will live within a hundred miles of the coasts.

The lifestyle of the Perpetual Beach Walker thus offers a second solution for ecological nomads. Optimize time in your comfort zone and maximize ecological security within this push-pull situation. In the push of wintertime, escape snowy blizzards, slush-filled boots, ice storms breaking utility lines and knocking out your electrical power. In your southern coastal abode, bright, sunny days unfold to pleasant temperatures from November through March. When the seasons change and it seems too hot in the South, follow the spring north, leaving behind the storm season of supercharged Greenhouse-world hurricanes.

The Time-Share Timer

Some Boomers need that psychological anchor, *one* home filled with the family treasures and photo albums. Paradoxically, the stronger the anchor holds them to one spot, the stronger grows their wanderlust. The emotional imperative to get out of town becomes almost overpowering. Their attitude is strongly voiced: "Once a month, come hell or high water, we're going to do something wildly fun, just for us. We always have a wild card. With a new trip every month to plan for and think about, we don't have more than two or three weeks to wait for it."

Some Boomers have crafted a unique lifestyle strategy tied to purchasing limited ownership of many separate places. The magic word is **time-share**. Think of your nest eggs split up among one big basket and then lots and lots of very little baskets. With one permanent base, you have one place to hold all your stuff. More importantly, this "real" home houses your washer and dryer—critical appliances to help you recover from the last trip and prepare for the next one.

Time-share condominiums offer a cost-effective way to diversify your kinds of lifestyle experiences and locations. You get a predictable quality and a modest purchase price. When you purchase a time-share, you join other buyers to become part owners of a particular condominium. You buy the rights to live in that furnished suite for the same two-week interval every year, for the rest of your life. By acquiring time-shares, you share the financial and ecological risk of owning property. Through time-share associations, you can trade your two-week window in any given year to vacation in another place and another two-week block of time.[14]

Boomers can invest in part-time ownership of a *new time-share each year* of the next decade. That adds up to a different time-share to enjoy for two weeks of each month nearly all year long. By the Boomer Breakpoint 2010, the lifestyle of the **Time-Share Timer** can guarantee that seasonal round of lifestyle options so necessary for the well-traveled ecological nomad. The key to this strategy is to plan well for what you want in a Greenhouse world. Factor in how climate warming will affect the kinds of things you want to see and do. For example, in the Great Smoky Mountains, the timing will change for peak wildflower viewing. The Wildflower Pilgrimage may be gradually rescheduled from late April much earlier into March. Prime time for autumn shows of brilliant leaf color will be progressively later in the southern Appalachian Mountains, moving from middle October well into November. On the Gulf and Atlantic Coasts, peak hurricane activity occurs today during the first week of September. In coming decades, the amount of this hurricane activity should increase, and the peak will broaden through late September and into early October. The strategy for timing your time-shares is not to choose the most expensive prime times right now. Rather, choose the less expensive "shoulder seasons" *that will become* the most desirable prime times for your retirement future.

The Sun-Seeking Road Warrior

The concept of time-sharing has expanded from the bricks and mortar of permanent condominiums to the impermanence of an asphalt drive-through. Increasingly popular as a Boomer strategy is life spent literally on the road as the **Sun-Seeking Road Warrior**. Recreational vehicles, or RVs, are self-contained homes on wheels with all the creature comforts. You can wake up on the road to morning coffee and Danish rolls, use the cellular phone, plug in to cable television, and get up-to-date stock market quotes via satellite. With location readouts available from global positioning satellites (the GPS system), you'll know precisely where you're at, even if you're lost. The migratory lifestyle as a road warrior offers freedom, serendipity, and yet, if you wish, a seasonal round-trip through all your favorite places. Of course, burning fossil fuel to power up the RV is not environmentally benign, but then again, think of all the heating oil or other fuel you're not consuming to stave off winter in the frozen northland or the electricity you're not using to air-condition a large home in the sweltering southland.

Mobile-oriented time-share resorts for RVs offer parking space for sale and utility hookups for weeks, months, or the season, all guaranteed for life. You select your own itinerary of ecological safe sites. For example, if riding horses is your life's passion, hook up the horse trailer to the back of your RV. Follow the grass greening in the pastures as the growing season unfolds. Head out for wide open spaces and networks of riding trails. Spend a month at a time at suitable campgrounds, enjoy the spontaneous mix of fellow RVers over evening campfires and barbecues.

Major organizations like the Good Sam Club, with more than a million RV members, provide a clearinghouse for social interactions and rendezvous points for forming up new caravans of ecological nomads.[15] With instant access via satellite dish to the Weather Channel, travel routes can readily be changed to focus on the good weather or to escape bad storms.

This strategy of one RV for all places might be your preference, tailored to optimize quality of life while minimizing financial and ecological risk.

The Seafaring Vagabond

Remember that famous line, "one if by land, or two if by sea," from Henry Wadsworth Longfellow's epic poem, *The Midnight Ride of Paul Revere*? Those two lanterns, hung in the belfry of the Old North Church in Boston, served as powerful beacons that sparked Paul Revere's tumultuous ride to warn fellow patriots of the approaching march by the British.[16]

These days, some Boomers are watching very different kinds of beacons—navigation beacons sent out by lighthouse, radio signal, or satellite for boating enthusiasts to home in on.[17] For mariners, the phrase "two if by sea" now conjures up a very creative strategy for the ecological nomad— two yacht berths if by sea.

Seafaring Vagabonds follow the winds and their fancy, swept away by the tides. This lifestyle option favors serendipity in discovering picturesque ports of call. Yet predictably, these maritime drifters still may be drawn to two seasonal anchoring sites, two dockside berths for their yachts or houseboats in convenient winter and summer marinas. Aquatic nomads will succeed *well ahead of the Wave and above the Surge*. They've closed that office door back in the city, slipped out of town ahead of the Age Wave of retiring Boomers, and settled into a new floating reality.

It's said a rising tide lifts all boats. With rising sea levels in a Greenhouse world (figure 4-5), these vagabonds stay floating above the hurricane's storm surge. The mariner's simple solution is to buy a longer mooring line to tie up to the dock. If you live on a houseboat in Key West and a hurricane threatens, just sail into the sunset and away from the storm. If all else fails and you have to flee inland and the storm sinks your boat, no problem. Buy another one. This may seem a cavalier attitude, but as your insurance company well knows, catastrophes happen. The lifestyle of a boating Ecological Nomad offers maximum enjoyment of the here and now, mobility for wherever your quest leads you, and only a modest level of ecological risk. This adaptive solution to the Greenhouse world potentially yields high reward for only intermediate risk.

Each of the five mobile lifestyle strategies can assure that comfort seekers will live well in a Greenhouse world. As an Ecological Nomad, you create a seasonal round of excursions from one home place to another home place or time-share condominium. Or, as you travel by RV on land or by yacht on the sea, you take the whole home place along for the ride. However, the flip side is this: if financial constraints or personal decisions tether you permanently to only one geographic location, you've lost precious mobility. When Greenhouse-world disasters threaten, your only option then is to hunker down and persist as an **Ecological Survivor** (table 9-1).

Stability Seekers: Year-Rounders

In an unsure and rapidly changing world, the often illusory combination of *very high reward for virtually no ecological risk* is the lifetime desire for the **Year-Rounders**, which we also call the **Stability Seekers**. Paradoxically, the all-pervasive fixation of many Boomers on *financial* risk effectively blindsides them to *ecological* risk, the specter of Greenhouse-world changes unfolding before us all. This dichotomy quickly separates out the Greenhouse winners, year-rounders who plan well and capture the stability they seek. Unfortunately, the year-rounders who guess wrong or simply react to catastrophes after the fact are destined to become **Ecological Losers**, left behind in the dust (table 9-1).

The Stability Seekers deliberately choose one of several possible Greenhouse-world strategies that avoid disturbance, maximize predictability, and ensure ecological security. These conservative visionaries plan their legacy for the very long term. Year-Rounders plan to put down roots in one location, an ecological safe site, establishing bonds with their community and with their regional landscape. If you belong to this group, your specific selection of place may tie to your ancestral heritage, perhaps the original pioneer homestead settled four or five generations back. Alternatively, rather than re-establish your American family roots in the old place, you may elect to migrate once again to make a fresh start.

The Homesteader

For some Boomers, variety in experience rather than place is truly the spice of life. A predictable annual round of four distinctly different seasons offers the diversity of lifestyle activities we welcome. In the next decades, you may have to relocate northward to continue enjoying winter snowfall. Yes, but where to move? Milder winters will characterize the region spreading from the Pacific Coast of Washington and British Columbia across the Upper Midwest and Great Lakes to New England and the maritime provinces of eastern Canada. However, to enjoy thick and persistent snowpack, remember the routes taken consistently by major midwinter storms (figure 5-3). Then opt for retirement locations adjacent to extensive bodies of water, such as the Great Lakes or along the northern Pacific coastline or the Atlantic Seaboard. For example, the zone of serious lake-effect snowfall will develop along the storm track of Alberta Clippers and downwind of the northernmost Great Lakes. A winter wonderland in the Greenhouse world awaits you along the southern and eastern shores of Lake Superior and northern Lakes Michigan, Huron, and Ontario (figure 5-2). Energetic nor'easter storms should funnel snow-producing gales into the St. Lawrence Valley as well as the maritime sectors of Maine, New Brunswick, Nova Scotia, and Newfoundland. More frequent winter storms of glaze ice, however, will cripple electrical utilities and road traffic in more temperate portions of the southern Great Lakes and southern New England.

The early "shoulder season" of spring will start sooner and persist longer. Tornado Alley in the Greenhouse world will expand with springtime storms that will swarm across the Great Plains into central Canada and race through the western Great Lakes states of Minnesota, Wisconsin, and Lower Michigan. The chilly surface waters of Lakes Michigan, Superior, and Huron will provide a natural protective shield, shutting down northeastward-moving storms and choking off the tornadoes potentially spawned along the squall line. Northern Michigan and south-central Ontario offer a natural, sheltered haven from assault by these springtime tornadoes.

Your regional trade-offs for the summer season will be tied to the inevitable Greenhouse-world increase in numbers of hot and extremely hot days. Summer warmth will extend northward, baking midwestern cities increasingly prone to infrastructure degradation and electrical power brownouts. Cooler summer temperatures will continue to prevail within the "sea breeze zones of convection" bordering the Great Lakes and the oceans. Southern Canada will provide the buffered mild climate sought by heat-stressed resorters.

Good advice for Year-Rounders in the northland who build that ulti-mate lakeside bungalow may be to disregard the sage wisdom of your build-ing contractor! Design for passive solar heating and cooling. Be sure to add in that heat-pump system in order to ensure your cooling comfort in the Greenhouse-world summers that lie ahead. Insulate and orient your new home to let in breezes and screen out summer heat. Consider renewable energy systems to generate your own electricity. Recurring droughts will intensify in severity, producing new dust bowls across the northern Great Plains as well as in landlocked interiors of states bordering the Great Lakes. Catchwater systems that harvest rainwater, and graywater systems that recycle and reuse some wastewater, can make a lot of sense. One more bit of advice—spend extra money for digging that water well even deeper than your building contractor thinks is necessary. At least you'll have grim satis-faction when your cost-conscious neighbors have shallower wells go dry.

The autumn season will arrive later and persist longer, well into what has traditionally been the window of wintertime for states bordering Canada. September and October will bring hurricanes up the Atlantic Seaboard as far as New York State. Their drenching rains will saturate the soil, and the high sustained winds of these energetic hurricanes will topple mature trees across New England before the storms slam into the northern Appalachian Mountains. The Greenhouse-world combination of milder, wetter winters, much hotter and drier summers, and more variable spring and fall seasons bringing more tornadoes and hurricanes, respectively, will threaten the very survival of forest species as well as the out-of-doors lifestyle pursuits of many Boomer Year-Rounders. The Greenhouse-world scenarios project wholesale forest diebacks and extinctions of many important tree species in the eastern United States (figures 5-5 and 5-6). The climatic shelter offered by lake-affected coasts (figure 5-2) will buffer Greenhouse-world stresses for Great Lake forests. The topographic terrain of the northern Appalachians will also help to preserve the biological diversity of natural plant and animal life.

What's the bottom line? The Year-Rounder who wants to maximize both quality of life and ecological security *in one place* can enjoy *four seasons*, environmental protection from droughts, tornadoes, and hurricanes, and long-term preservation of biologically rich natural landscapes. The geographic regions of potentially suitable locations for the lifestyle of the **Homesteader** lie somewhere along the eastern shores of Lakes Superior, Michigan, Huron, and Ontario, around smaller lakes of the central and eastern Great Lakes region, and eastward to the mountain crests of the northern Appalachians. The strategy of one place for all seasons can be successful, for example, in states such as Michigan and New York, and in the southern provinces of Ontario and Quebec.

The Sustainable Hedonist

The ultimate strategy for many Stability Seekers may be to discover one location for a dream home with a constant climate year-round, both today and in the future Greenhouse world. This ultraconservative option for the **Sustainable Hedonist** is challenging, blending high quality of life with low ecological risk. If your ideal comfort zone lies in the subtropical to tropical range, equatorial regions may be the appropriate destination for you. Economic futurist and demographer Harry S. Dent, Jr., champions the prospective Boomer paradise of the Caribbean Islands, where he plans to retire.[18] This is a reasonable choice for one place, with a fairly consistent temperature range year-round, but the Caribbean setting carries with it considerable ecological risk. The magnitude and frequency of hurricanes will greatly accelerate in the new Greenhouse world reality. Prospective destinations within the Caribbean Sea or that border the Gulf of Mexico have a very big sign erected on them—a sign that says "Hurricane, hit me!" or, if you prefer, "Bam! Bam!"

A far better option for one place with one equable season that will last even into the Greenhouse world can be found in our fiftieth state, on the island archipelago of Hawaii. The EPA summary for Greenhouse conditions in Hawaii projects only a 3°F increase in overall temperatures by the year 2100, naturally buffered by the vast waters of the Pacific Ocean surrounding these islands.[19] Changes in precipitation are the wild card here—El Niño and La Niña will dictate Greenhouse-world weather patterns. Hawaii's subtropical weather may shift toward greater seasonal variation in rainfall, and the islands will have greater exposure to passing tropical storms forming in the equatorial Pacific. They will, however, remain a tropical paradise.

Winners in a Greenhouse World

As Boomers look forward to realizing vibrant new lifestyles after retirement, our quest for living well, with the highest possible quality of life, may involve a succession of personal-decision breakpoints. To cope with them successfully, we can employ a lifelong series of strategies shaped by our changing tolerance to ecological risk (table 9-1). As we reach our fifties and sixties, we may look for big opportunities created in high-risk areas by Greenhouse-world change. In our seventies, we may elect to abandon the high-risk high wire of Thrill Seekers to seek comfort instead as Ecological Nomads. In our eighties and beyond, we may require greater security and stability, consolidating all our resources as Year-Rounders. But right now, as we enter the twenty-first century, Boomer Breakpoints await us, marking our journey ahead as Greenhouse winners.

CHAPTER

Legacy for
Future Generations

Our neighbors in Knoxville own eighty acres of mostly forested hills and hollows in rural Sevier County, east Tennessee. From their property near the headwaters of Walden Creek, a nearly continuous series of interconnecting ridges can be traced to the westernmost boundary of the Great Smoky Mountains National Park. The forested hillsides serve as an extension of habitat for wide-ranging large mammals such as black bear. North-facing hillslopes along Walden Creek harbor dense rhododendron arbors that shelter the showy blooms of white and nodding trilliums, Dutchman's breeches, and hepaticas. Another suite of spring ephemeral wildflowers including lemon-flowered trilliums, wild geraniums, and trout lilies grace the complementary south-facing slopes.

Our friends have a traditional Appalachian log cabin, with squared wood beams interlocked by dovetailed corners. Their weekend retreat lies nestled on a flat terrace above spring flood stage, located in a secluded hollow in which still grow gnarled apple trees that were planted around the first pioneer homesteads in this valley, dating back five generations.

But just over the next ridge, in nearby Wears Valley located between Walden Creek and the Great Smoky Mountains National Park, changes have been taking place recently that have alarmed our friends. The expanding tourist industry has greatly increased demand for rental cabins and retirement homes with glorious views of the high mountain peaks. Developers have stripped large sections of land of their trees, leaving red earthen scars eroding downslope and causing both landslides and slumping hillslopes. The swift pace of progress is rapidly transforming this pastoral setting from a mosaic of grassy valleys and forested hillsides, characteristic of much of both Sevier and Knox counties in the early to middle twentieth century, to a scarred vestige of that countryside now falling victim to urban and suburban sprawl. As new subdivisions spring up all over the region, areas that were formerly continuous forest become increasingly fragmented, with homesites developed on cleared parcels of one to ten acres. Today the verdant old-growth forests of the Great Smoky Mountains National Park are increasingly being isolated as neighboring forests are cut down. The park boundary contrasts sharply with the adjacent artificial landscape of amusement parks and shopping malls of the tourist towns Gatlinburg and Pigeon Forge, Tennessee, and the casinos of Cherokee, North Carolina.

Our friends near Walden Creek worry that the continuing development will eliminate much of the natural biological diversity unique to the Smoky Mountains region. Their earnest wish is for their natural wildflower gardens to be perpetuated long after they have turned over the property to their children and grandchildren. During their lifetimes their hideaway is secure, but for longer protection of the biologically diverse and vulnerable hillsides, they have turned to a local conservation organization. Our neighbors have placed all of the land except for the mowed, flat stream terrace where their homestead is located into a land trust with the Foothills Conservancy.[1] As an **ecological set-aside**, their hidden valley will be protected forever, with no possibility that the forest will be clear-cut or that the hillsides will be commercially developed. In this way, our good friends can regain peace of mind and truly leave a lasting legacy for future generations.

No Park Is an Island: Corridors and Buffer Zones

Joni Mitchell's song is haunting: "They paved paradise, put up a parking lot." Why is this simple act such an irreversible travesty, particularly in a Greenhouse world? The answer has to do with corridors for plant and animal migration and with ecological buffer zones around national parks, as well as between publicly owned land and private developments.

Ecologists Susan Bratton and Jonathan Ambrose have studied the effects of increasing the habitat contrast between natural areas and developed areas.[2] Along a series of transects across the boundaries of the Great Smoky Mountains National Park, they found that songbird diversity diminished dramatically across the park border and into the city of Gatlinburg, and that the numbers of nonnative species of plants and animals increased dramatically along the edge of the park relative to its interior. Suburbanization and urbanization have profound effects not only on the appearance of the landscape but also on its ecological functioning.[3] As the natural vegetation such as forest is fragmented into ever smaller parcels, fewer native species can continue to live within the shrinking remnants of the former large ecosystems. The first species of animals to be eliminated are those with large territories, such as black bear and wolf. The next species to go are those that require sizable areas of "interior" vegetation, such as closed forest, and that cannot tolerate environmental conditions at the margins of natural vegetation. Suburban landscapes generally have very low biological diversity of native plant and animal species, except for those very hardy species that are adapted to edges between forests and fields, for example white-tailed deer and blackberry brambles. Such sites are, however, readily invaded by nonnative weeds.[4]

In order to maintain high biological diversity of native species, sufficient area and extent of native environments must be deliberately planned for and preserved. One way to accomplish this is to set aside large, contiguous reserves of land, called **biosphere reserves**, in national parks. In the literature on conservation biology, much debate has centered on whether it is better to set aside such large, self-contained, island-like preserves to retain the highest possible diversity of habitats and species, or whether it is better in the long run to establish a larger network of smaller reserves interconnected by extended greenway corridors. Known as the **SLOSS (Single Large or Several Small) reserves dilemma**, this predicament arises from the higher probability that native species confined to one "island" will go extinct because of restricted interbreeding than if several different populations are able to intermingle and reproduce along interconnected corridors of suitable habitat. In a Greenhouse world the problem will become acute, because changing climate may shift the location of optimal living requirements for many rare species "out from under" the fixed location of an island-like reserve (figure 10-1).[5]

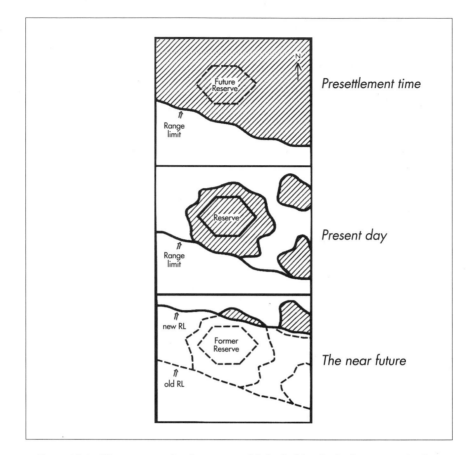

Figure 10-1. **Climate-warming impacts on biological/ecological reserves.** As the Greenhouse-world environment changes, the range of habitat suitable for many species might "migrate" right out of the reserve established to protect them. (Adapted from R. L. Peters and J. D. S. Darling, "The Greenhouse Effect and Nature Reserves," *BioScience* 35 [1985]:707–717.)

But meanwhile progress marches on, heedless of this debate. Increasingly, biosphere reserves such as the Smokies are becoming more and more like isolated islands in a figurative sea of developed landscape. The more isolated the Smokies become, the more vulnerable to extinction will be the species they harbor, especially in the face of Greenhouse-world climate change.[6] To counteract this tendency toward isolation, the biosphere reserve system has been developed within an overall design consisting of a **core area,** where human activities are limited and where maintaining wildlife habitat and biological diversity are the primary goals, plus a **buffer zone** that surrounds the core area. In the buffer zone, increasing amounts of human activity are allowed, with less activity near core areas and increased residential, commercial, and industrial development away from core areas. But many species of wildlife are also supported because sufficient areas of native vegetation are preserved to allow for contiguity of habitat. **Corridors** between core-area reserves are an essential part of this conservation plan, particularly as they allow for genetic exchange among a species' populations and also allow for migrations of species as plant hardiness zones begin to shift in the near future.[7]

With these concerns and basic principles for establishing biological reserves in mind, the **Wildlands Project** has been established as a nationwide effort to coordinate professional biologists, biological diversity activists, rural communities, and land use planners to coordinate the linking of biological reserves across the continent of North America.[8] Such a continent-wide conservation network will include core wilderness areas connected by corridors and surrounded by buffer zones. The overall objective of the Wildlands Project is to keep some places wild but also to provide a less destructive way for people and natural biological communities to co-exist. The Wildlands Project is broad in scope, with the long-term intent to link biosphere reserves in Canada, the United States, Mexico, and Central America. It has two major focuses at present, in the northern Rocky Mountains and across peninsular Florida.[9]

What Can the Concerned Citizen Do?

"Think globally, act locally." This slogan is now well worn, but still useful. As we have seen from our Greenhouse lifestyle chart and cross-impacts analysis, many aspects of global climate change are beyond the influence of the average person. Regulation of global emissions of atmospheric pollutants is a matter for multinational agreements, made on the level of the United Nations, the World Trade Organization, and the heads of the economically powerful G-7 countries.[10] Many of the climate change documents cited in this book are directed at such government officials, produced as information that will be useful in slowing or abating some of the deleterious worldwide effects of global climate change.

So what can an individual citizen do? We need to remember David Gates's advice.[11] While it is "already too late" to prevent global warming, it is "never too late" to become educated about the possible consequences and to make informed decisions about how to cope with or adapt to the new Greenhouse-world conditions. For most of us, this means acting locally, in our own communities.

As concerned citizens, we can make a difference on several levels. We must prepare to cope effectively with catastrophic change. Our first task is to recognize the changes that are likely to occur around our own homesites and to prepare for such environmental hazards as hurricanes, floods, and tornadoes with appropriate renovation and adequate insurance coverage. Choosing environmentally friendly lifestyles and being conscientious consumers will minimize our own impacts that would otherwise contribute further to climate change. We can discourage alien plant species from taking over our properties and encourage the survival of native songbirds by landscaping with native plants, not only in our own backyards but also along golf course greens and in public parks.[12]

On a broader scale, we can participate in community planning to preserve wetlands that are a source of biological diversity, to maintain substantial areas of remnant natural vegetation, to ensure the survival of threatened species of native plants and animals, and to help communities targeted by waves of retiring Baby Boomers to cope with looming population pressures. Finally, if we own a sizable tract of forest, wetland, or mountainside, we can donate or sell the property to a conservancy program, or we can have it set aside in a land trust to remove it from future commercial development. This conservancy option carries with it considerable tax advantages in addition to the peace of mind obtained from being a responsible steward of the land.

The Nature Conservancy

The Nature Conservancy, like any organization, is simply a group of people working together to do something they cannot do individually.

Victoria M. Edwards

The Nature Conservancy was founded in 1951 as an offshoot of the Ecological Society of America, a national organization of professional academic and government ecologists. The mission of The Nature Conservancy is "to preserve plants, animals, and natural communities that represent the diversity of life on Earth by protecting the lands and water they need to survive."[13] With some 1,500 preserves covering more than one million acres and containing 1,750 rare species and natural communities, it is the largest private system of nature sanctuaries in the world. Some three-fourths of the annual donations to The Nature Conservancy are made by individual contributors. All donations, whether cash or real estate, are fully tax deductible. The Nature Conservancy maintains a Natural Heritage Inventory Program, a network of natural area inventories designed and established with assistance from the conservancy but administered by state departments of natural resources. Through the natural area inventories, critical habitats for preserving rare species (including plants, mammals, and migrating birds) are identified for purchase and preservation. The stewardship program of The Nature Conservancy is concerned with restoring habitat as well as monitoring and protecting species diversity, not only on conservancy-owned land, but on private and public land as well.

Land trusts, jewels of the living landscape

America's fastest-growing conservation group is the whole collection of organizations known as **land trusts**. Originally conceived in nineteenth-century Massachusetts as a means of preserving areas of historic interest or natural beauty, today over one thousand land trusts exist in the United States as local, regional, or statewide nonprofit organizations dedicated to protecting important areas of private land.[14] Most land trusts are funded by membership fees and donations, and over three million people are involved in this effort today. In addition, private foundations help to fund **green space**.

In contrast to The Nature Conservancy, which focuses on biological diversity at the national and global level, land trusts reflect the values of the local community. For example, in east Tennessee the Foothills Conservancy has developed a plan to protect a number of large tracts of land totaling about twenty-seven thousand acres in the vicinity of the Great Smoky Mountains National Park. The conservancy envisions this endeavor as creating a buffer zone between the national park and the greater Knoxville metropolitan area, thus protecting both wildlife and the "viewshed" of the national park.

Local citizens, such as our friends on Walden Creek, are encouraged to develop a **conservation easement,** a legal agreement between landowner and land trust in which landowners, without giving up ownership of their property, voluntarily restrict the type and amount of development that may take place on it. In this way, our friends can sell their land or leave it to their heirs, but future owners will be legally bound by the terms of the easement agreement. The immediate benefit to landowners is that the value of the easement is treated as a charitable gift and therefore is deductible from income tax. One of the long-term benefits of this type of agreement is the peace of mind that comes from knowing you are making a difference in helping to sustain biological diversity. Secretary of the U.S. Department of the Interior Bruce Babbitt has called these visionary plans of the Foothills Conservancy a "model for the nation."[15]

A second example of land trusts in which the average citizen can become directly involved is the Little Traverse Conservancy, located in northwestern Lower Michigan. The stated purpose of the Little Traverse Conservancy is "to protect the natural diversity and beauty of Northern Michigan by preserving significant natural land and scenic areas, and to foster appreciation and understanding of the environment. . . . The Conservancy is a broad coalition of individuals, families, and businesses who agree that the acquisition and protection of natural land is important if we are to retain the quality of life which makes northern Michigan so attractive. The Little Traverse Conservancy is supported entirely by people who willingly donate their time, talent, and financial support to protect irreplaceable natural land."[16] This conservancy is not a lobby group. Rather than engage in political activism, the organization seeks to protect scenic views, open spaces, and wildlife habitat to benefit not only biological but also local human communities; all conservancy-owned land is open to the public. The Little Traverse Conservancy grows both by land acquisition and conservation easements, and has an active environmental education program.

For each of these examples, both in the Great Smoky Mountains and on lakeshores of northern Lower Michigan, land trusts enhance the long-term desirability of local communities as retirement destinations. They help to ensure an ecologically sustainable future for their regions as green communities develop within natural settings.

Green communities

Marine biologist Jane Lubchenco, in her role as president of the Ecological Society of America, worked with a committee of professional ecologists to outline the urgency of acting to sustain biological diversity into the future.[17] In her presidential address to the American Association for the Advancement of Science in 1998, Lubchenco took this concept a step further, calling for a *new social contract for science* that will "exercise good judgement, wisdom, and humility and help society move toward a more sustainable biosphere."[18] How can we translate this science-based social contract into deliberate actions that Boomers and others can take?

Landscape ecologist Joan Iverson Nassauer seeks to bridge the gap between what suburbanites and wildlife biologists can do by inventing new landscapes that not only embody ecological function but also accommodate human needs. Nassauer has designed a model three-hundred-acre subdivision for the city of Cambridge, Minnesota, located in the expanding "commuter-shed" of the 2.5-million-population Twin Cities metropolis of Minneapolis–St. Paul. Nassauer did *not* use the default solution. She rejected the conventional plan for suburb development intended to maintain an area's rural character. That would have placed each individual home in the middle of a cleared ten acre land parcel, with homes lining the shores of two lakes within the development. Instead, she designed an alternative plan *for a green community,* creating a form of subdivision that is both familiar and radically new.

Nassauer developed this innovative approach to landscape design in order "to be a source of pride, to look well cared for, to create a sense of ownership, to look safe, to be legible, to afford prospect and refuge, to create a feeling of closeness to nature . . . to include affordable housing, to be accessible by public transportation, to provide public access to high amenity landscape features, and to minimize infrastructure costs." Other goals, important for the environment but not necessarily immediately apparent to home buyers or developers, were to improve surface and groundwater quality and to increase habitat quality, connectivity, and extent. Thus, this ecologically sustainable subdivision was designed with biological diversity and conservation goals as well as cost-effective and energy-efficient cultural means.

Nassauer's overall plan (figure 10-2) clusters residences in one sector of the subdivision, with an extended woodland adjoining the clustered homes.

Two restored wetlands connect surface water drainage with nearby lakes. A large upland prairie and meadow remain intact adjacent to the lakes, and an extensive park and jogging trail offer a safe area for exercise and recreation in a natural setting.[19]

Another example of the need for ecologically correct landscape planning can be found in the coastal zone of southeastern Georgia. The Golden Isles are a series of barrier islands that today harbor about two-thirds of all the maritime live oak forest that remains from its previously extensive habitat, which stretched from the Carolinas to northern Florida. The United States Department of Interior estimates that at the present rate of development of the coastal zone, fully half of the maritime forest that still exists will be developed by the year 2005.[20]

Loss of this vulnerable habitat will have a cascading series of effects on wildlife. For example, during seasonal bird migrations, maritime forests are important for the Cape May warbler, black-throated warbler, the rare Kirtland's warbler, and Connecticut warbler, as well as the regionally scarce painted bunting. Resident birds eat live oak acorns as well as berries from coastal shrubs including yaupon and wax myrtle. Spanish moss, a member of the pineapple family that typically drapes in festoons from the low branches of live oak trees, is a nesting material for osprey and provides nesting habitat for a variety of other birds.

Spanish moss creates a distinctive, vertically oriented microhabitat within the maritime forest, which supports dozens of species of arthropods and insects, both within the live oak branches and in the ground litter. Within the coastal zone where this natural, biologically diverse community was once widespread, subdivisions typically have partitioned the forest into small lots of one-fourth to one-half acre in size. This means that the forest canopy is becoming fragmented into noncontiguous units that are too small to sustain the diversity of life in the original forest community.

Future human community development in the maritime live oak forest habitat needs to consider both the extent of road networks and the orientation of roads. Any road or trail corridors that are opened through the forest allow salt-laden winds to be funneled through the forest, killing native tree and shrub species and allowing alien species to become established. Just constructing roads in the usual way, like bowling alley swaths cut straight

Ecological Corridor Neighborhood Design Plan

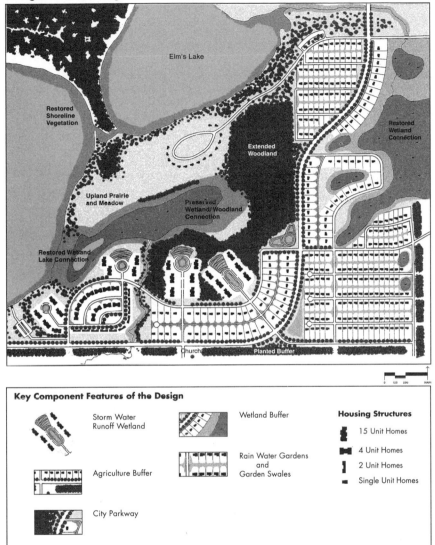

Elm's Lake

Restored
Shoreline
Vegetation

Restored
Wetland
Connection

Extended
Woodland

Upland Prairie
and Meadow

Preserved
Wetland/Woodland
Connection

Restored Wetland
Lake Connection

Church   Planted Buffer

Key Component Features of the Design

Storm Water
Runoff Wetland

Wetland Buffer

Housing Structures

15 Unit Homes

4 Unit Homes

Agriculture Buffer

Rain Water Gardens
and
Garden Swales

2 Unit Homes

Single Unit Homes

City Parkway

Figure 10-2. **Subdivision plan for Cambridge, Minnesota, a "green" community.** This subdivision, designed to put landscape-ecological principles into practice, may become a model for future development. (Adapted from Joan Iverson Nassauer, "Culture as a Means for Experimentation and Action," pages 129–133, in John A. Wiens and Michael R. Moss, eds., *Issues in Landscape Ecology* [Snowmass Village, Colo.: International Association for Landscape Ecology, Fifth World Congress, 1999].)

down to the beach, can further degrade the natural biological diversity of these sensitive ecosystems.

The idea of sustainable development is catching on with Boomers and others. The EPA Green Communities Program is dedicated to developing **sustainable green communities,** that is, communities that "integrate a healthy environment, a vibrant economy, and a high quality of life."[21] For communities in the southern Appalachian Mountains, a group called Southern Appalachian Man and the Biosphere (SAMAB) is developing a major Sustainable Communities Initiative to provide critical information on ecological, economic, and demographic characterization of counties and watersheds, to be used as indicators of long-term sustainability. This planning tool (available on the Internet) offers individuals, nonprofit organizations, and government planning units easy access to sources of detailed information relevant to sustainable development.[22]

In South Carolina, upper-end resort communities being developed on barrier islands such as Dewees Island are beginning to advertise that "development and environment are natural allies . . . because building in harmony with the environment is less expensive than dominating or destroying natural resources." One of the advertised advantages of "green living" on Dewees Island is that "homes are designed to 'nest' within the habitat, taking advantage of winter/summer sun, shade, prevailing breezes and natural lighting . . . only indigenous or native vegetation to the South Carolina coastal plain are allowed . . . removing the need for irrigation, fertilizers, and pesticides."[23]

Planning resort developments for long-term ecological and cultural sustainability, however, takes more than configuring for present-day resources. As we have outlined in earlier chapters, an additional consideration is planning for environmental variability that will accompany global climate change. Barrier islands such as Dewees, Bald Head, and other resort areas of the Carolinas will be susceptible to an ever increasing threat from category 4 and category 5 hurricanes. In order to ensure that your beach home will not only be integrated (maybe even assimilated) within the existing environment but will become a legacy you can pass on to the next generation, planning with an awareness of the likelihood of storm surge damage or shoreline erosion is a smart strategy.

The Ecological Edge

In this chapter, we have emphasized the ways that we boomers can leave a positive legacy for the future. As we stated in our preface, most of us fall into a middle ground between extreme lifestyles. Generally speaking, Boomers are not overly greedy capitalists bent on acquiring ever more materialistic possessions. Nor are most of us likely to return to our hippie roots and become residents of isolated, out-of-sync communes. On the whole, we have worked hard to build our careers, to gain gender and racial equality in the workplace, and to afford a comfortable lifestyle. We are attuned to the need to make a difference, and we hope to live long enough to enjoy the fruits of our labors. Those of us who are centrists are neither extreme risk takers nor complacent about our place in the world. We are looking ahead to choosing retirement lifestyles that are rich in amenities but not at the expense of environmentally and socially responsible living environments.

We have suggested here two ways that Boomers can make a difference in the Greenhouse world—what we call gaining the **Ecological Edge** for the future. One way that our legacy for the future can be secured is by helping to maintain biological diversity, by becoming stewards of the land. We can donate time, land, or dollars to organizations such as land trusts and The Nature Conservancy. In so doing, we contribute to the creation of biological preserves and the development of interconnected corridors that will assist plants and animals to survive the coming changes in climate. We can work on a more local level by landscaping our own surroundings with native plant species, around our homes and yards and even along fairways on our golf courses. In this way, we can create habitats for native birds and wildlife, and at the same time enrich our lives by their presence.

A second fundamental way in which Boomers can influence the course of events as we retire in droves to seaside, sunbelt, mountain, or lakeside resort destinations is to take an active role in community planning. We can minimize our collective impact on already established communities and help to enhance the environment for all who live in those destination areas. We can do this through planning green communities that emphasize both ecological and cultural sustainability.

Sustainability is a key word in making a better future in the Greenhouse world. We may not be able to control the ultimate course of events globally, but we can contribute meaningfully to our local and regional well-being. Living well in the Greenhouse world means much more than owning yachts, sports cars, and mansions by the sea. To truly live well, we must all come to grips with what it means to be responsible stewards of the land and to build sustainable communities that will be the foundation of a worthy legacy for future generations.

We conclude by reiterating the need for considered and reasoned action based on enlightened self-interest. Readers of this book, and users of our Web site, can take advantage of opportunities to set themselves up for the good life based on their awareness of the realities of climate change. At the same time they can take specific actions to contribute to community sustainability and to preserve and restore their local environments. If we acknowledge that in the twenty-first century we face major decisions about lifestyle changes, and that their timing will coincide with major Greenhouse-world environmental changes, we can become better prepared to face the coming Boomer Breakpoints with creatively adaptive solutions and contingency plans based on the best available eco-futuristic scenarios. We can use our Ecological Survival Kit to adapt and to optimize successful strategies as Greenhouse winners. We can then use our ecological edge to live well in the Greenhouse world, and to have peace of mind in helping to preserve our legacy for future generations.

Notes

Chapter 2. Baby Boomers

1. William Sterling and Stephen Waite, *Boomernomics: The Future of Your Money in the Upcoming Generational Warfare* (New York: Library of Contemporary Thought, Ballantine Books, 1998).

2. Landon Y. Jones, *Great Expectations: America and the Baby Boom Generation* (New York: Ballantine Books, 1980).

3. The Baby Boomer Web site is compiled by the Population Division of the United States Census Bureau. This electronic document can be viewed online at http://www.census.gov/population/www/socdemo/age.html/. This site itemizes the vital statistics for Baby Boomers, with national and state summaries for social and economic profiles.

4. Ken Dychtwald, *Age Power: How the 21st Century will be Ruled by the New Old* (New York: Jeremy P. Tarcher/Putnam, 1999).

5. Harry S. Dent, Jr., *The Great Boom Ahead: Your Comprehensive Guide to Personal and Business Profit in the New Era of Prosperity* (New York: Hyperion, 1993); Harry S. Dent, Jr., *The Roaring 2000s: Building the Wealth and Lifestyle You Desire in the Greatest Boom in History* (New York: Simon and Schuster, 1998); Harry S. Dent, Jr., *The Roaring 2000s Investor: Strategies for the Life You Want* (New York: Simon and Schuster Trade, 1999). Harry S. Dent's Web site is at http://www.hsdent.com/.

6. The U.S. Social Security Administration maintains actuarial life tables (see Life Table for 1997) and gender-specific and age-group profiles

for the population insured with retirement benefits guaranteed by Social Security trust funds. For more details, click on its Web site at http://www.ssa.gov/OACT/STATS/statTab.html/.

7. See Harry S. Dent's characterizations cited in note 5.

8. William Gates and C. Hemingway, *Business @ the Speed of Thought: Using a Digital Nervous System* (New York: Warner Books, 1998).

9. For the Boomer Initiative Web site, see http://www.babyboomers.com/.

10. Jon Gertner, "What is Wealth? More Americans that ever enjoy six-figure incomes, and some are reaching even higher. The members of the new *ultra* middle class—riding a tidal wave of prosperity—are living their own, customized American dreams." *Money,* vol. 29, no. 13 (December 2000): 94–107; David Brooks, *Bobos in Paradise: The New Upper Class and How They Got There.* New York: Simon & Schuster, 2000.

11. Paul H. Ray and Sherry Ruth Anderson, *The Cultural Creatives: How 50 Million People Are Changing the World* (New York: Harmony Books, 2000).

12. For a profile of the "Cultural Creatives," describing how their values and lifestyles influence consumer purchases of $230 billion in the United States and $546 billion worldwide, read three magazine articles: Paul Ray, "Who Is the LOHAS Consumer?" on pages 35–38, 52; the interview with Paul Ray written by Nancy Nachman-Hunt, "Discovering the Cultural Creatives: Market research guru Paul Ray speaks out on how he found them and why you should care" on pages 39–40, 52; and Monica Emerich, "LOHAS Means Business. The $230 billion U.S. Lifestyles of Health and Sustainability Marketplace has 50 million consumers waiting to buy into it" on pages 32–34. All three articles are published in the March/April 2000 issue (vol. 1, no. 1) of *Natural Business LOHAS Journal, Tracking the Lifestyles of Health and Sustainability Market.* The LOHAS (http://www.LohasJournal.com/) targets five consumer segments: ecological lifestyles, sustainable economy, healthy lifestyles, alternative health care, and personal development. To participate in the Cultural Creatives Web ring, check out the online magazine and chat room sponsored by ecoPlanet's EcoChoices Community (http://ecochoicescommunity.com/).

13. Alvin Toffler, *Future Shock* (New York: Random House, 1970).

14. See Sterling and Waite's *Boomernomics* for more on this strategy (note 1).

15. Street-savvy ex-cop John Corey expresses this crusty view of life on page 163 of Nelson DeMille's novel, *The Lion's Game* (New York: Warner Books, 2000).

16. Daniel McFadden, "Demographics, the Housing Market, and the Welfare of the Elderly," in David A. Wise, ed., *Studies in the Economics of Aging*, a National Bureau of Economic Research Project Report. (Chicago: University of Chicago Press, 1994); N. Ravo, "Will the Roof Cave In? After sifting through supply, demand, and demographics, one economist sees the real-estate market slipping down hill into the next century," *Worth*, vol. 8., no. 2 (1999), 73–75.

17. Sterling and Waite, *Boomernomics* (see note 1).

18. Sylvester J. Schieber and John B. Shoven, eds., *The Consequences of Population Aging on Private Pension Fund Saving and Asset Markets*. Center for Economic Policy Research, Publication no. 363 (Palo Alto: Stanford University, 1993). Sylvester J. Schieber and John B. Shoven, eds., *Public Policy toward Pensions: A Twentieth Century Fund Book* (Cambridge: The MIT Press, 1997).

19. See the Social Security Administration's position paper entitled *The Future of Social Security* (Pub. no. 05-10055), that was published in April 2000. This electronic document can be viewed online at http://www.ssa.gov/pubs/10055.html/.

20. Schieber and Shoven, eds., *Consequences of Population Aging* (see note 17).

Chapter 3. Greenhouse Warming—Somebody Else's Problem?

1. John T. Houghton, *Global Warming: The Complete Briefing*, 2d ed. (New York: Cambridge University Press, 1997). Sir John Houghton's account of his Greenhouse briefing with Prime Minister Thatcher appears on pages xi–xii and 144. Check out the IPCC Web site (http://www.ipcc.ch/about/about.htm/) for a wealth of information on the scientific and political options for dealing with global climate change.

2. David M. Gates, *Climate Change and Its Biological Consequences* (Sunderland, Mass.: Sinauer Associates, 1993). See also Houghton, *Global Warming* (note 1).

3. Stephen H. Schneider, *Global Warming: Are We Entering the Greenhouse Century?* (San Francisco: Sierra Club Books, 1989). This book provides

a rare eyewitness account of the intriguing personalities who developed the new computer models that simulate future climate change, and the fierce scientific and political debates they sparked.

4. Schneider, *Global Warming*, gives more detail (see note 3).

5. Gates, *Climate Change* (see note 2).

6. Houghton, *Global Warming* (see note 1). For access to comprehensive data sets and current analyses for Greenhouse gases, check out the CDIAC Web site for atmospheric trace gases at http://cdiac.esd.ornl.gov/about/intro.html/.

7. For an elegant yet easily understandable statement on the linkage between CFCs and ozone, read the chapter "Back from Beyond the Limits" in the book by Donella H. Meadows, Dennis L. Meadows, and Jørgen Randers, *Beyond the Limits: Confronting Global Collapse, Envisioning a Sustainable Future* (White River Junction, Vt.: Chelsea Green, 1992). This bestseller explores futuristic models for the population explosion and vulnerable environments. Recent geopolitical action on the ozone problem, or "ozone hole," represents the first successful global effort to recognize and then solve an imminent ecological crisis. For a current report on this continuing international collaboration, check out the status of "Ozone Depletion" as documented on the Web by the United Nations' Global Environmental Facility (http://www.undp.org/gef/new/).

8. Houghton, *Global Warming*; Gates, *Climate Change*; Schneider, *Global Warming* (see notes 1–3).

9. Raymond S. Bradley, *Paleoclimatology: Reconstructing Climates of the Quaternary*, 2d ed., International Geophysics Series, vol. 64 (London: Harcourt/Academic Press, 1999); Tom M. L. Wigley, M. J. Ingram, and G. Farmer, *Climate and History: Studies in Past Climates and Their Impact on Man* (Cambridge: Cambridge University Press, 1985); John F. Fialka, "U.S. Study on Global Warming May Overplay Dire Side," *The Wall Street Journal* (May 26, 2000), p. A24.

10. Kwang.-Y. Kim and Thomas J. Crowley, "Modeling the Climate Effect of Unrestricted Greenhouse Emissions over the Next 10,000 Years," *Geophysical Research Letters* 21(1994): 681–684.

11. Schneider, *Global Warming* (see note 3); and John T. Houghton, L. G. Meira Filho, B. A. Callander, N. Harris, A. Kattenberg, and K. Maskell,

Climate Change 1995: The Science of Climate Change (Cambridge: Cambridge University Press, 1996).

12. Schneider, *Global Warming* (see note 3).

13. Houghton, *Global Warming* (see note 1).

14. Tom M. L. Wigley, *The Science of Climatic Change: Global and U.S. Perspectives* (Arlington, Va.: PEW Center of Global Change, 1999). This report is available online at http://www.pewclimate.org/projects/env_science.html/.

15. Jim E. Hansen, A. Lacis, D. Rind, G. Russell, P. Stone, I. Fung, R. Ruedy, and J. Lerner, "Climate Sensitivity: Analysis of Feedback Mechanisms," pages 130–163, in Jim E. Hansen and T. Takahashi, eds., *Climate Processes and Climate Sensitivity*, Maurice Ewing Series, 5 (Washington, D.C.: American Geophysical Union, 1984). GISS results on global climate modeling can be found online at http://www.giss.nasa.gov/data/. For short popular summaries concerning GISS model predictions of climate changes, check the science reports on http://www.giss.nasa.gov/rescarch/intro/.

16. Richard T. Wetherald and Syokuro Manabe, "An Investigation of Cloud Cover Change in Response to Thermal Forcing," *Climatic Change* 8 (1986): 5–23; Syokuro Manabe and Richard T. Wetherald, "Large-Scale Changes of Soil Wetness Induced by an Increase in Atmospheric Carbon Dioxide," *Journal of the Atmospheric Sciences* 44 (1987): 1211–1235. For current GFDL results, check out the following Web sites: http://www.gfdl.gov/gfdl_research.html/; http://www.ncdc.noaa.gov/ol/climate/online/gcm.html/; and http://www.ncdc.noaa.gov/gcps/gcps.html/. Effective GFDL Web sites for user-friendly stories on weather, climate, and Greenhouse climatic change are located at http://www.gfdl.gov/~gth/web_page/climate_and_weather.html/.

17. Schneider, *Global Warming* (see note 3).

18. Houghton, *Global Warming*, p. 92 (see note 1).

19. Meadows et al., *Beyond the Limits* (see note 8).

20. J. A. Leggett, W. J. Pepper, and R. J. Swart. "Emissions Scenarios for the IPCC: An Update," pages 69–95, in John T. Houghton, B. A. Callander, and S. K. Varney, eds., *Climate Change 1992: The Supplementary Report to the IPCC Scientific Assessment* (Cambridge: Cambridge University Press, 1992).

21. Tom M. L. Wigley and S. C. B. Raper, "Global Mean Temperature and Sea Level Projections under the 1992 IPCC Emissions Scenarios," pages

401–404, in R. A. Warrick, E. M. Barrow, and T. M. L. Wigley, eds., *Climate and Sea Level Change: Observations, Projections, and Implications* (Cambridge: Cambridge University Press, 1993).

22. Gates, *Climate Change* (see note 2).

23. For an intriguing overview of relevant and effective corporate actions in reducing Greenhouse-gas emissions, eco-activists should read the timely report written by Robert R. Nordhaus and Stephen C. Fotis, *Early Actions and Global Climate Change: An Analysis of Early Action Crediting Proposals* (Arlington, Va.: PEW Center for Global Change, 1999), found online at http://www.pewclimate.org/projects/pol_early.html/ .

Chapter 4. The Seaside

1. Several major Greenhouse-world syntheses provide estimates of population growth anticipated along U.S. coastlines: T. J. Cuilliton, M. Warren, T. Goodspeed, D. Rerner, C. Blackwell, and J. McDonough, III, *Fifty Years of Population Change along the Nation's Coasts: 1960–2010* (Rockville, Md.: National Oceanic and Atmospheric Administration [NOAA], 1990); M. L. Miller and J. Auyong, "Coastal Zone Tourism: A Potent Force Affecting Environment and Society," *Marine Policy* (1991), pages 75–99; Rhode Island Sea Grant, Coastal Ocean Office, *NOAA's Coastal Ocean Program: Science for Solutions* (Washington, D.C.: NOAA, 1992); James E. Newmann, Gary Yohe, Robert Nicholls, and Michelle Manion, *Sea Level Rise and Global Climate Change: A Review of Impacts to U.S. Coasts* (Arlington, Va.: PEW Center for Global Climate Change, 2000), available online at http://www.pewclimate.org/projects/env_sealevel.pdf/ (this electronic publication contains maps showing the spatial pattern of how much sea level will rise by the years 2010 and 2030, inundating coastal zones of the conterminous United States); J. C. Field, D. F. Boesch, D. Scavia, R. Buddemeier, V. R. Burkett, D. Cayan, M. Fogarty, M. Harwell, R. Howarth, C. Mason, R. A. Park, L. J. Pietrafesa, D. Reed, T. Royer, A. Sallenger, M. Spranger, and J. G. Titus, "Potential Consequences of Climate Variability and Change on Coastal Areas and Marine Resources," chapter 16 in the report compiled by the U.S. National Assessment Team (NAST), *Climatic Change Impacts for the United States: The Potential Consequences of Climatic Variability and Change, Foundation Document* (Washington, D.C.: U.S. Global

Change Research Program [USGCRP], 2000), available at http://www. gcrio.org/NationalAssessment/foundation.html/.

2. V. Gornitz, "Mean Sea Level Changes in the Recent Past," pages 25–44, in R. A. Warrick, E. M. Barrow, and T. M. L. Wigley, eds., *Climate and Sea Level Change: Observations, Projections, and Implications* (Cambridge: Cambridge University Press, 1993).

3. Thomas M. L. Wigley and S. C. B. Raper, "Global Mean Temperature and Sea Level Projections under the 1992 IPCC Emissions Scenarios," pages 401–404, in Warrick, Barrow, and Wigley, eds., *Climate and Sea Level Change* (see note 2); and S. C. B. Raper, "Observational Data on the Relationships between Climate Change and the Frequency and Magnitude of Severe Tropical Storms," pages 192–212 in the same volume.

4. For more information on EPA projections of future sea-level rise, see the Web site at http://www.epa.gov/globalwarming/climate/future/sealevel.html/.

5. Joel B. Smith, Rich Richels, and Barbara Miller, "Potential Consequences of Climatic Variability and Change for the Western United States," chapter 8 in the report compiled by the U.S. National Assessment Synthesis Team (NAST), *Climatic Change Impacts for the United States: The Potential Consequences of Climatic Variability and Change, Foundation Document* (Washington, D.C.: U.S. Global Change Research Program [USGCRP], 2000), with both the summary Overview and the detailed foundation documents of the report available at http://www.gcrio.org/NationalAssessment/ foundation.html/.

6. J. W. Day, W. H. Conner, R. Costanza, G. P. Kemp, and I. A. Mendelssohn, "Impacts of Sea Level Rise on Coastal Systems with Special Emphasis on the Mississippi River Deltaic Plain," pages 276–296, in Warrick, Barrow, and Wigley, eds., *Climate and Sea Level Change* (see note 2).

7. James Titus and Vijay Narayanan, *The Probability of Sea Level Rise* (Washington, D.C.: Environmental Protection Agency [EPA], 1995). For an online version of this report (#EPA 230-R95-008), see the Web site at http:// www.epa.gov/globalwarming/publications/impacts/sealevel/probability/ chapt_9.pdf/.

8. R. A. Warrick, "Climate and Sea Level Change: A Synthesis," pages 3–21, in Warrick, Barrow, and Wigley, eds., *Climate and Sea Level Change*; also Wigley and Raper, "Global Mean Temperature and Sea Level Projections," in the same volume (see note 3).

9. For more details on the Saffir-Simpson hurricane scale, see the Web site of the National Hurricane Center at http://www.nhc.noaa.gov/aboutsshs.html/.

10. See the Web site at http://www.nhc.noaa.gov/ for more details and the monitoring of such storms as they form and move in a given hurricane season.

11. See the Web site at http://www.nhc.noaa.gov/pastint.html/.

12. See the Web site at http://www.fema.gov/library/strmfrm.htm/. Provided by the Federal Emergency Management Agency (FEMA), this Web site has graphs, courtesy of the Disaster Archives, showing both total numbers and average numbers per year of tropical storms forming in the Equatorial and North Atlantic Oceans, tallied by week during the hurricane season. The pronounced peak of hurricane formation occurs during the first week of each September.

13. See the National Hurricane Center Web site (note 9) for these guidelines.

14. See the Web site for the Miami-Dade Building Code for South Florida construction at http://www. buildingcodeonline.com/ordi1.htm/.

15. See the references in note 3.

16. Raper, "Observational Data" (see note 3).

17. See the Web site at http://www.giss.nasa.gov/research/intro/druyan.03/.

18 K. A. Emanuel, "The Dependence of Hurricane Intensity on Climate," *Nature* 326 (1987): 483–485; other data from M. E. Schlesinger, "Model Projections of CO_2-induced Equilibrium Climate Change," pages 169–191, in Warrick, Barrow, and Wigley, eds., *Climate and Sea Level Change* (see note 3).

19. W. K. Michener, E. R. Blood, K. L. Bildstein, M. M. Brinson, and L. R. Gardner, "Climate Change, Hurricanes and Tropical Storms, and Rising Sea Level in Coastal Wetlands," *Ecological Applications* vol. 7, no. 3 (1997): 770–801.

20. This historic record of hurricanes in the North Atlantic Ocean (1899–1992 used by Michener et al. [note 19]) is admittedly not as long as we would like. However, it provides a useful empirical basis for calculating the recurrence interval, or typical return time, between successive hurricane landfalls along the Atlantic and Gulf Coasts.

21. Note that the historic record of Michener et al. (note 19) extends back only to the year 1899, and that two major hurricanes made landfall in southeastern Georgia in years just preceding the beginning of record keeping, in 1893 and 1898.

22. Newmann, Yohe, Nicholls, and Manion, *Sea Level Rise and Global Climate Change* (see note 1).

23. Tom M. L. Wigley, *The Science of Climate Change: Global and U.S. Perspectives* (Arlington, Va.: PEW Center for Global Change, 1999), with the online publication available at the Web site http://www.pewclimate. org/projects/env_science.html/. Page 33 of this report discusses how Greenhouse hurricanes will carry more moisture and dump greater rainfall amounts along American coasts.

24. Charles C. Coutant, "Temperature-Oxygen Habitat for Freshwater and Coastal Striped Bass in a Changing Climate," *Transactions of the American Fisheries Society* 119 (1990): 240–253.

25. See the Web site at http://www.fema.gov/nfip/laws.htm/.

26. See the Web site at http://www.fema.gov/nfip/readme.htm/.

27. Federal Emergency Management Agency (FEMA), "44 CFR Part 61, RIN 3067-AD02. National Flood Insurance Program (NFIP); Insurance Coverage and Rates. Proposed Rules; 'Target Repetitive Loss' Buildings," *Federal Register* vol. 64, no. 150 (August 5, 1999), pages 42, 632–42, 633 (see online Web site, note 25).

28. James G. Titus, *Greenhouse Effect, Sea Level Rise, and Coastal Wetlands* (Washington, D.C.: U.S. Environmental Protection Agency, 1988); James G. Titus, "Greenhouse Effect and Coastal Wetland Policy: How Americans Could Abandon an Area the Size of Massachusetts at Minimum Cost," *Environmental Management* 15 (1991): 39–58 (available online at http://www.epa.gov/globalwarming/publications/impacts/sealevel/massachusetts.pdf/); James G. Titus, "Rising Seas, Coastal Erosion, and the Takings Clause: How to Save Wetlands and Beaches without Hurting Property Owners," *Maryland Law Review* vol. 57, no. 4 (1998): 1279–1399 (available online at http://www.epa.gov/globalwarming/publications/impacts/sealevel/takings.pdf/); D. S. Shriner, R. B. Street, R. Ball, D. D'Amours, K. Duncan, D. Kaiser, A. Maarouf, L. Mortsch, P. Mulholland, R. Neilson, J. A. Patz, J. D. Scheraga, J. G. Titus, H. Vaughan, and M. Weltz, "North America," pages 253–330, in R. T. Watson, M. C. Zinyowera, and R. H. Moss, eds.,

The Regional Impacts of Climate Change: An Assessment of Vulnerability, Special Report of IPCC Working Group II (Cambridge: Cambridge University Press, 1998), available online at http://www.epa.gov/globalwarming/publications/reference/ipcc/chp8/america.html/.

29. Cornelia Dean, *Against the Tide: The Battle for America's Beaches* (New York: Columbia University Press, 1999).

30. See the Web site at note 26.

31. See notes 28 and 29.

32. Emanuel, "Hurricane Intensity" (see note 18).

33. For convenient summaries of "coastal ecological characterizations" and maps showing historic shoreline changes, see the many Coastal Resource Atlases prepared for the National Coastal Ecosystems Team, in connection with the Office of Biological Services, Fish and Wildlife Service, U.S. Department of Interior. Check out the literature citations for "other Biological Reports" at the Web site http://www.nwrc.gov/publications/pubslist.html/. For example, for the southern Atlantic Seaboard, look at the Coastal Resource Atlas prepared by Jane Davis. The full reference is Jane S. Davis, M. D. McKenzie, J. V. Miglarese, R. H. Dunlap, J. J. Manzi, and L. A. Barclay, eds., *Ecological Characterization of the Sea Island Coastal Region of South Carolina and Georgia, Resource Atlas, Publication FWS/OBS-79/43* (Washington, D.C.: jointly published by the National Coastal Ecosystems Team, Office of Biological Services, Fish and Wildlife Service, U.S. Department of Interior, and for the Office of Research and Development, U.S. Environmental Protection Agency, 1980).

Chapter 5. The Lakeshore

1. See the Web site of the Mackinac Bridge Authority at http://www.mackinacbridge.org/nav.html/ for a view of current weather conditions and traffic flow on Interstate 75 over the Mackinac Bridge, connecting the Upper and Lower Peninsulas of Michigan.

2. M. E. Schlesinger, "Model Projections of CO_2-induced Equilibrium Climate Change," pages 169–191, in R. A. Warrick, E. M. Barrow, and T. M. L. Wigley, eds., *Climate and Sea Level Change: Observations, Projections, and Implications* (Cambridge: Cambridge University Press, 1993). In this paper, Schlesinger summarizes the computer results from five separate General Circulation Models (GCMs) with the simulated dynamics of the

atmosphere coupled to the oceans, driven by the seasonal cycle of the sun's energy. Computer simulation runs contrasted Greenhouse and present-day conditions, focusing on environmental changes imposed by doubling atmospheric levels of carbon dioxide relative to today's value. The results portray the magnitude of change anticipated for air temperatures at the land surface for wintertime and for summertime. Separate "difference maps" for soil moisture show regional Greenhouse-world patterns of moisture availability, as future precipitation levels are reduced.

Current GFDL results can be seen online at the Web site http://www.gfdl. gov./gfdl_research.html/. The maps of Greenhouse-world climate based on the GISS model can be viewed at the Web site http://www.giss.nasa. gov/data/.

3. As senior scientist at the National Center for Atmospheric Research, Tom Wigley has gathered together the Greenhouse-climate results simulated by fifteen GCMs. Averaging these computer results into a composite map, Wigley can see how well these various climate models agree and where their forecasts diverge in regional scenarios. This approach presents big-picture Greenhouse-world maps for the United States, contrasting the regional differences in temperature and moisture, expressed as Greenhouse-world departures from today's "normal" pattern. For atmospheric conditions with a doubling in concentration of carbon dioxide over preindustrial levels, seasonal patterns of average temperature and precipitation are mapped across the lower forty-eight states for winter, spring, summer, and fall. Showing spatially detailed results of climate standardized to incremental amounts (each 1°C, or about 1.5°F) of global mean warming, Greenhouse-world maps can be produced for various projected times in the near future. Based on the amount of warming expected for North America, Greenhouse-world maps can be produced for Boomer Breakpoints 2010, 2025, and 2070. View these extraordinary maps of Greenhouse-world climate in Wigley's 1999 PEW report at the Web site http://www.pewclimate.org/projects/ env_science.html/.

4. The National Assessment offers one of the best current statements about anticipated Greenhouse-world impacts for the Great Lakes region. Check out this electronic publication: David R. Easterling and Thomas R. Karl, "Potential Consequences of Climate Variability and Change for the Midwestern United States," chapter 6 in the report compiled by the U.S.

National Assessment Team (NAST), *Climatic Change Impacts for the United States: The Potential Consequences of Climatic Variability and Change, Foundation Document* (Washington, D.C.: U.S. Global Change Research Program [USGCRP], 2000), available at the Web site http://www.gcrio.org/ NationalAssessment/foundation.html/. For more information, see the Executive summary on "Adapting to the Impacts of Climate Change and Variability," prepared as a Workshop Report by the Great Lakes–St. Lawrence Basin (GLSLB) Project, a multi-year project of the Canadian Federal Green Plan Initiative. See the Web site at http://www.on.ec.gc.ca/glimr/metadata/ adapting-climate/intro.html/. The Web site at http://www.on.ec.gc.ca/glimr/ search.html/ offers a powerful search engine for scientific matters related to "Our Great Lakes: Working Towards a Healthy and Sustainable Great Lakes Basin Ecosystem," sponsored by Environment Canada.

5. Henry G. Hengeveld, "Global Climate Change: Implications for Air Temperature and Water Supply in Canada," *Transactions of the American Fisheries Society* 119 (1990): 176–182.

6. Hengeveld, "Global Climate Change" (see note 5); and Easterling and Karl, "Potential consequences" (see note 4).

7. Easterling and Karl, "Potential Consequences" (see note 4); and C. A. Crissman, "Impacts of Electricity Generation in New York State," in U.S. National Climate Change Program Office and the Canadian Climate Centre, *Impacts of Climate Change in the Great Lakes Basin*, Joint Report no. 1 (Oak Brook, Ill.: National Climate Change Program, 1988).

8. T. E. Crowley, II, and H. C. Hartmann, "Effects of Climate Change on the Laurentian Great Lake Levels," pages 4-1 to 4-34, in J. B. Smith and D. A. Tirpak, eds., *The Potential Effects of Global Climate on the United States, Appendix A, Water Resources*, EPA-230-05-89-051 (Washington, D.C.: U.S. Environmental Protection Agency [EPA], 1989).

9. U.S. Environmental Protection Agency, *Ecological Impacts from Climate Change: An Economic Analysis of Freshwater Recreational Fishing*, EPA Report 220-R-95-004 (Washington, D.C.: EPA, 1995). In rivers and streams, cold-water fish are most likely to be affected as Greenhouse warming alters their thermal habitat, particularly along the southernmost border of their modern distributional range. For an online abstract of this EPA report, click on http://www.epa.gov/globalwarming/publications/impacts/ eco_fishing.html/.

10. John J. Magnuson, J. Donald Meisner, and David K. Hill, "Potential Changes in the Thermal Habitat of Great Lakes Fish after Global Climate Warming," *Transactions of the American Fisheries Society* 119 (1990): 254–264; David K. Hill and John J. Magnuson, "Potential Effects of Global Climate Warming on the Growth and Prey Consumption of Great Lakes Fish," *Transactions of the American Fisheries Society* 119 (1990): 265–275.

11. Michael J. McCormick, "Potential Changes in Thermal Structure and Cycle of Lake Michigan due to Global Warming," *Transactions of the American Fisheries Society* 119 (1990): 183–194.

12. Magnuson, Meisner, and Hill, "Potential Changes" (see note 10).

13. John J. Magnuson, D. K. Hill, H. A. Regier, J. A. Holmes, J. D. Meisner, and B. J. Shuter, "Potential Responses of Great Lakes Fishes and Their Habitat to Global Climate Warming," pages 2-1 to 2-42, in Smith and Tirpak, eds., *Potential Effects* (see note 8).

14. Hill and Magnuson, "Potential Effects" (see note 10).

15. Magnuson, Meisner, and Hill, "Potential Changes" (see note 10).

16. D. C. Norton and S. J. Bolsenga, "Spatiotemporal Trends in Lake Effect and Continental Snowfall in the Laurentian Great Lakes, 1951–1980," *Journal of Climate* 6 (1993): 1943–1956.

17. Van L. Eichenlaub, *Weather and Climate of the Great Lakes Region* (Notre Dame, Ind.: The University of Notre Dame Press, 1979); Van L. Eichenlaub, J. R. Harman, F. V. Nurnberger, and H. J. Stolle, *The Climate Atlas of Michigan* (Notre Dame, Ind.: The University of Notre Dame Press, 1990).

18. S. N. Rodionov, "Association between Winter Precipitation and Water Level Fluctuations in the Great Lakes and Atmospheric Circulation Patterns," *Journal of Climate* 7 (1994): 1693–1706.

19. See note 4.

20. To learn more about our approach to forest history and the stories behind our scientific discoveries, you may want to see Hazel's nontechnical book: Hazel R. Delcourt, *Reading Landscapes of the Past: Pleistocene Refuges, Postglacial Emergence, and Future of the Eastern Deciduous Forest* (Granville, Ohio: McDonald and Woodward Publishing Company, 2001). For links to this publication as well as other related books, see our Web site at http://www.boomerbreakpoints.com/.

21. To gain an impression of how much North American landscapes have changed just in the past few centuries, since the first European colonists arrived, see the following nontechnical treatment of forest and landscape history by the current director of the Harvard Forest: David R. Foster, *Thoreau's Country: Journey through a Transformed Landscape* (Cambridge, Mass.: Harvard University Press, 1999).

22. An excellent source of information about the various techniques for deciphering landscape history is the book by Emily W. B. Russell, *People and the Land through Time: Linking Ecology and History* (New Haven, Conn.: Yale University Press, 1997).

23. See note 20 and also Hazel R. Delcourt and Paul A. Delcourt, *Quaternary Ecology, a Paleoecological Perspective* (New York: Chapman and Hall, 1991).

24. F. I. Woodward, "Review of the Effects of Climate on Vegetation: Ranges, Competition, and Composition," pages 105–123, in R. L. Peters and T. E. Lovejoy, eds., *Global Warming and Biological Diversity* (New Haven, Conn.: Yale University Press, 1992).

25. Robert S. Thompson, Katherine H. Anderson, and Patrick J. Bartlein, *Atlas of Relations between Climatic Parameters and Distributions of Important Trees and Shrubs in North America,* vol. 1, *Introduction and Conifers* (paper 1650-A), and vol. 2, *Hardwoods* (paper 1650-B), *U.S. Geological Survey (USGS) Professional Paper* 1650-A and -B (Denver: USGS, 1999). Paper copies are available for purchase from the U.S. Geological Survey, Information Services, P.O. Box 25286, Federal Center, Denver, CO 80225. Online versions of these publications are available at the Web site http://greenwood.cr.usgs.gov/pub/ppapers/p1650-a/ and at http://.greenwood.cr.usgs.gov/pub/ppapers/p1650-b/.

26. H. M. Cathey, *USDA Plant Hardiness Zone Map* (map scale 1:6,000,000; geographic coverage of Canada, the United States, Mexico, Hawaii, and the Aleutian Islands). U.S. Department of Agriculture, Agricultural Research Service, Miscellaneous Publication no. 1475 (Washington, D.C.: USDA, 1990). Contemporary maps show eleven zones representing different bioclimatic regions of winter minimum temperatures that determine the important agricultural crops and the ornamental landscape plants that can be grown in each area. Cold-hardiness ratings are available for selected woody plants, and representative species are indicated by corre-

sponding hardiness zone. Available from the U.S. National Arboretum, the 1998 Web version of the map of plant hardiness zones can be viewed online at http://www.ars-grin.gov/na/hardines.htm/. Close-ups of any geographic area can be brought to screen using the hyperlink.

27. Linda Joyce, John Aber, Steve McNurly, Virginia Dale, Andrew Hansen, Lloyd Irland, Ron Neilson, and Kenneth Skog, "Potential Consequences of Climate Variability and Change for the Forests of the United States," chapter 17 in the report compiled by the U.S. National Assessment Team (NAST), *Climatic Change* (see note 4).

28. We have mapped the directions and rates of spread for major trees of eastern North America as they moved northward from southern ice-age refuges beginning as early as 17,000 years ago. For a more detailed discussion of the interpretation of fossil evidence, see our book: Paul A. Delcourt and Hazel R. Delcourt, *Long-Term Forest Dynamics of the Temperate Zone*, Ecological Studies 63 (New York: Springer-Verlag, 1987). Our collaborative syntheses of forest response to long-term climate change across North America can be viewed on Jonathan Adams's Web site showing vegetation maps for the last 150,000 years (http://www.esd.ornl.gov/projects/qen/nercNORTHAMERICA.html/).

29. See note 28.

30. Margaret B. Davis and Catherine Zabinski, "Changes in Geographic Range Resulting from Greenhouse Warming: Effects on Biodiversity in Forests," pages 297–308, in Peters and Lovejoy, eds., *Global Warming* (see note 24).

31. Based on his life's work in plant ecology, Dan Botkin has proposed a thoughtful program for managing nature, a nature we Boomers will literally mold with our actions in the Greenhouse world. We recommend the following book as required reading for eco-literacy: Daniel B. Botkin, *Discordant Harmonies: A New Ecology for the Twenty-first Century* (Oxford: Oxford University Press, 1990).

32. Daniel B. Botkin and R. A. Nisbet, "Projecting the Effects of Climate Change on Biological Diversity in Forests," pages 277–293, in Peters and Lovejoy, eds., *Global Warming* (see note 24).

33. See the reference in note 32, and Daniel B. Botkin, "Global Warming and Forests of the Great Lakes States: An Example of the Use of Quantitative Projections in Policy Analysis," pages 154–166, in J. Schmandt and

J. Clarkson, eds., *The Regions and Global Warming: Impacts and Response Strategies* (Oxford: Oxford University Press, 1992).

34. Botkin, "Global Warming and Forests" (see note 33).

35. Botkin and Nisbet, "Projecting the Effects" (see note 32).

36. Botkin, "Global Warming and Forests" (see note 33).

37. See the publications of Jerry Franklin, an eminent forest ecologist who studies sustainable forestry practices on landscapes in the Pacific Northwest: Jerry F. Franklin, "Sustainability of Managed Temperate Forest Ecosystems," pages 355–385, in Mohan Munasinghe and Walter Shearer, eds., *Defining and Measuring Sustainability: The Biogeophysical Foundations* (Washington, D.C.: The International Bank for Reconstruction and Development/The World Bank, 1995); Jerry F. Franklin, "The Natural, the Clearcut, and the Future," *Northwest Science* vol. 72, no. 2 (1998): 134–138; Jerry F. Franklin, Frederick J. Swanson, Mark E. Harmon, et al., "Effects of Global Climatic Change on Forests in Northwestern North America," *The Northwest Environmental Journal* 7 (1991): 233–254; Jerry F. Franklin, "A kinder, gentler forestry in Our Future: the Rise of Alternative Forestry," *The Trumpeter* vol. 6, no. 3 (1989): 99–100. A listing of Franklin's publications is available online from the H. J. Andrews Experimental Forest, a National Science Foundation long-term ecological research site located in the western Cascade Range, Blue River, Oregon at this Web address: http://www.fsl.orst.edu/lter/homepage.htm/.

38. The Wildland Fire Management Program, within the Forest Management Division of Michigan's Department of National Resources (DNR), maintains the Web site for its Mushroom Hunting Guide at http://www.dnr.state.mi.us/www/fmd/fire/MushroomReport_1.html/. The list identifies burned areas larger than ten acres, located by county (with map coordinates of Section, Township, and Range) and by vegetation cover type. Wild morel mushrooms are typically found growing in hardwood forests under the canopy of deciduous trees; shrubby thickets of wild blueberries are most commonly discovered on recently burned sites in evergreen conifer forests and in boggy wetlands.

Chapter 6. The Mountains

1. Horace Kephart, *Our Southern Highlanders: A Narrative of Adventure in the Southern Appalachians and a Study of Life among the Mountaineers* (Knoxville: University of Tennessee Press, 1977).

2. Great Smoky Mountains National Park, *Restoration of Native Brook Trout* (Gatlinburg, Tenn.: Great Smoky Mountains National Park, 1999). This report can be viewed online at the Web site http://www.nps.gov/grsm/brtrout.htm/; R. Houk, *A Natural History Guide: Great Smoky Mountains National Park* (New York: Houghton Mifflin, 1993). To see an acid-rain map with contemporary pH levels for precipitation, check out the Web page of the United States Geological Survey (USGS) at http://www.water.usgs.gov/nwc/NWC/pH/html/ph.html/.

3. C. M. Vanhooser, "To Protect and Preserve: an Unlikely Coalition of Officials and Volunteers Work Side by Side to Save the Smokies' Aquatic Treasures." *Southern Living* vol. 33, no. 2 (1998): TL-8, 10, 12.

4. Southern Appalachian Assessment (SAA), *Atmospheric Technical Report*, no. 3 (Knoxville, Tenn.: Southern Appalachian Man and Biosphere Cooperative [SAMAB], 1996). Five technical reports, including the overall summary and syntheses for aquatic life, atmospheric conditions, social/cultural/economic patterns, and terrestrial ecosystems are available online at the SAMAB Web site, http://samab.org/saa/reports/. Major data sets for geographic information on natural resources are coordinated through the Southern Appalachian Regional Information System (SARIS), available at http://samab.org/data/data.html/.

5. Jim R. Renfro, "Trouble in High Places: Ozone Pollution," *Discovering the Smokies, a Science Journal* vol. 2, no. 1 (1999): 15.

6. Jim R. Renfro, "Evaluating the Effects of Ozone on Plants of Great Smoky Mountains National Park," *Park Service* 9 (1989): 1–22.

7. See the Southern Appalachian Mountain Initiative (SAMI) Web sites at http://www.saminet.org/ and http://www.rff.org/proj_summaries/files/burtraw_south_appalachian.htm/.

8. SAA, Atmospheric Technical Report (see note 4).

9. Steve Nash, *Blue Ridge 2020: An Owner's Manual* (Chapel Hill: University of North Carolina Press, 1999). See Steve Nash's Web site at http://www.blueridge2020.com/.

10. Web pages on air quality for the Smoky Mountains spell out the major problems, with Park air pollution levels representing some of our country's highest atmospheric concentrations recorded for ozone, nitrogen, and sulfur (see http://www.nps.gov/grsm/airq.htm/ and http://www.nps.gov/grsm/brairq.htm/). The Look Rock Observation Tower (elev. 2,670 feet) offers camera coverage with thirty-minute updates, with panoramic views sweeping looking eastward from Mt. LeConte (elev. 6,593 feet) to Thunderhead Mountain (elev. 5,527 feet) in the Great Smoky Mountains National Park; see the Smokies cam Web site at http://www2.nature.nps.gov/ARD/parks/grsm/lookRockWeather.htm/. This Look Rock station monitors visibility distance, atmospheric ozone levels, and meteorological conditions. Comparable camera stations for air-quality monitoring can also be seen for Acadia, Big Bend, Canyonlands, and Yosemite National Parks (see http://www2.nature.nps.gov/ard/gas/parkCams.htm/). For comparable air-quality information for the Northern Appalachians, click on http://www.neci.st.unh.edu/index.html/. The New England Climate Initiative (NECI) represents a key clearinghouse for *New England's Changing Climate, Weather and Air Quality*, coordinated by the Climate Change Research Center, University of New Hampshire, Durham.

11. See note 10.

12. See note 10.

13. See the references in note 4.

14. Environmental Protection Agency (EPA), *Acid Deposition Standard Feasibility Study* (Washington, D.C.: EPA, 1995). For more details, check out the Web site at http://www.epa.gov/ardpublic/acidrain/effects/execsum.html/.

15. See the references in note 2.

16. D. Kirk, *Smoky Mountains Trout Fishing Guide*, revised and updated edition (Hillsborough, N.C.: Menasha Ridge Press, 1997); H. L. Lawrence, *The Fly Fisherman's Guide to the Great Smoky Mountains National Park* (Nashville, Tenn.: Cumberland House, 1998).

17. J. D. Meisner, "Effect of Climate Warming on the Southern Margins of the Native Range of Brook Trout, *Salvelinus fontinalis*," *Canadian Journal of Fisheries and Aquatic Sciences* 47 (1990):1067–1070.

18. See note 2.

19. See note 2, and Meisner, "Effect of Climate Warming" (note 17).

20. See the reference in note 3.

21. See Meisner, "Effect of Climate Warming" (note 17).

22. For more on the environmental pressures imposed by growing numbers of tourists, check out the Web site for the Great Smoky Mountains National Park http://www.nps.gov/grsm/.

23. More information about this internationally significant biological preserve can be found at the Web site for the Southern Appalachian Man and Biosphere Cooperative at http://samab/org/.

24. Smokies Science Brief, "Old-growth Update," *Discovering the Smokies, a Science Journal* vol. 2, no. 1 (1999): 16.

25. Ken Wise and Ron Petersen, *A Natural History of Mt. LeConte* (Knoxville: University of Tennessee Press, 1998); Paul A. Delcourt and Hazel R. Delcourt, "Paleoecological Insights on Conservation of Biodiversity: a Focus on Species, Ecosystems, and Landscapes," *Ecological Applications* vol. 8, no. 4 (1998): 921–934.

26. The Great Smoky Mountains National Park is the site for what is termed the "moonshot of biological science," the All Taxa Biodiversity Inventory (ATBI). The ATBI is a pioneering attempt to identify all living plants and animals at a given location. The program goals include developing a comprehensive list of life forms in the park, estimated to number 100,000 different species, by the Boomer Breakpoint year 2010. For more information, see the Web site at http://www.discoverlife.org/ as well as publications of the Great Smoky Mountains Natural History Association, 115 Park Headquarters Road, Gatlinburg, TN 37738. See also the 1999 Note "All Taxa Biodiversity Inventory" on page 16 of vol. 2, no. 1, in the publication *Discovering the Smokies, a Science Journal*. To tap into the current taxonomic database for plant and animal species living in the Great Smokies, examine the Park's species list on file at http://ice.ucdavis.edu/nps/sbypark.html/.

27. C. C. Campbell, A. J. Sharp, R. W. Hutson, and W. F. Hutson, *Great Smoky Mountain Wildflowers: When and Where to Find Them*, 5th ed. (Northbrook, Ill.: Windy Pines Publishing, 1995); Peter White, T. Condon, J. Rack, C. A. McCormick, P. Beaty, and K. Langdon, *Wildflowers of the Smokies* (Gatlinburg, Tenn.: Great Smoky Mountains Natural History Association, 1996); Arthur Stupka, *Trees, Shrubs, and Woody Vines of Great Smoky Mountains National Park* (Knoxville: University of Tennessee

Press, 1964). Edwin Way Teale, *North with the Spring: a Naturalist's Record of a 17,000 Mile Journey with the North American Spring* (New York: Dodd Mead, 1952).

28. The Spring Wildflower Pilgrimage takes place Thursday through Saturday on the third weekend of April in the Great Smoky Mountains National Park. Begun in 1950 by the late University of Tennessee Professor of Botany Aaron J. Sharp, this Appalachian celebration is sponsored each year by the Park, the University of Tennessee Department of Botany, the Gatlinburg Garden Club, the Southern Appalachian Botanical Society, and the Great Smoky Mountains Natural History Association. Information on the next Spring Wildflower Pilgrimage can be found online at http://www.goldsword.com/wildflower/pilgrimage.html/.

29. See the *Watching Wildflowers* Web page (which will be available as of 2001) for the Education Committee at the Southern Appalachian Man and the Biosphere (SAMAB) Web site found at http://samab.org/educ/watching-wildflowers/.

30. Nash, *Blue Ridge 2020* (see note 9).

31. Great Smoky Mountains National Park (GSMNP), *Summary of Forest Insect and Disease Impacts in the Great Smoky Mountains National Park* (Gatlinburg, Tenn.: GSMNP, 1998). This report is available online at the following Web site: http://www.nps.gov/grsm/brinsect.htm/.

32. For a comprehensive look at species, see the ATBI Web site (note 26).

33. For example, for the states of Tennessee and North Carolina that form the mutual boundary of the Great Smoky Mountains, log on to the Web site at http://www.epa.gov/globalwarming/impacts/stateimp/index.html/.

34. David J. Hicks and Brian F. Chabot, "Deciduous Forest," pages 257–277, in Brian F. Chabot and Hal A. Mooney, eds., *Physiological Ecology of North American Plant Communities* (New York: Chapman and Hall Publishers, 1985).

35. White et al., *Wildflowers* (see note 27).

36. See note 28, for the next pilgrimage.

37. Albert J. Meier, Susan P. Bratton, and D. C. Duffy, "Possible Ecological Mechanisms for Loss of Vernal-Herb Diversity in Logged Eastern Deciduous Forests," *Ecological Applications* vol. 5, no. 4 (1995): 935–946.

38. Edwin Way Teale certainly thought so. See his *Autumn across America: A Naturalist's Record of a 20,000-Mile Journey through the North American Autumn* (New York: Dodd, Mead, 1956).

39. Hicks and Chabot, "Deciduous Forest" (see note 34).

40. Louis R. Iverson and Anantha M. Prasad, "Predicting Abundance of 80 Tree Species Following Climate Change in the Eastern United States," *Ecological Monographs* vol. 68, no. 4 (1998): 465–485; Louis R. Iverson, Anantha M. Prasad, B. J. Hale, and E. K. Sutherland, *An Atlas of Current and Potential Future Distributions of Common Trees of the Eastern United States*, General Technical Report NE-265 (Delaware, Ohio: Northeastern Research Station, USDA Forest Service, 1999). An interactive electronic version of this atlas is available online at the Web site http://www.fs.fed.us/ne/delaware/atlas/index.html/. For all counties of the United States lying east of 100°W longitude, Iverson and Prasad examined the computerized Forest Inventories for all tree species and related them to environmental conditions of climate, topography, soils, and land-use history. These quantitative relationships of forest species and their modern environment requirements reflect the underlying ecological controls governing eighty important tree species, allowing prediction of future adjustments in their geographic distribution. This published paper, atlas, and interactive Web site portray both present ranges and near-future geographic shifts of tree species as well as locations for their major forest population centers, as projected under five scenarios of Greenhouse climate change including the GISS and GFDL scenarios.

41. White et al., *Wildflowers* (see note 27).

42. Paul and Hazel Delcourt, "Paleoecological Insights" (see note 25).

43. See the references in note 25.

44. See note 25.

Chapter 7. The Sunbelt

1. Gary F. McCracken, M. K. McCracken, and A. T. Vawter, "Genetic Structure in Migratory Populations of the Bat *Tadarida brasiliensis mexicana*," *Journal of Mammalogy* 75 (1994): 500–514.

2. E. Zwingle, "Ogallala Aquifer: Wellspring of the High Plains," *National Geographic* (March 1993), pp. 80–107.

3. E. G. Gutentag, F. J. Heimes, N. C. Krother, R. R. Luckey, and J. B. Weeks, *Geohydrology of the High Plains Aquifer in Parts of Colorado, Kansas, Nebraska, New Mexico, Oklahoma, South Dakota, Texas, and Wyoming*, United States Geological Survey, Professional Paper 1400-B (Washington, D.C.: USGS, 1984). The United States Geological Survey (USGS) posts two important Web sites describing water-level changes in the high plains Ogallala Aquifer, as well as the geologic and hydrologic characteristics of this resource. See http://water.usgs.gov/public/wid/FS_215-95/FS_215-95.pdf/ and http://www-ne.cr.usgs.gov/highplains/hpactivities.html/.

4. The stunning magnitude of this cultural and ecological crisis is spelled out in two critical chapters: L. A. Joyce, D. Ojima, G. A. Seielstad, R. Harriss, and J. Jackett, "Potential Consequences of Climate Variability and Change for the Great Plains," and Katherine Jacobs, D. Briane Adams, and Peter Gleick, "Potential Consequences of Climate Variability and Change for the Water Resources of the United States," chapters 7 and 14 in the report compiled by the U.S. National Assessment Team (NAST), *Climatic Change Impacts for the United States: The Potential Consequences of Climatic Variability and Change, Foundation Document* (Washington, D.C.: U.S. Global Change Research Program [USGCRP], 2000), available at the Web site http://www.gcrio.org/NationalAssessment/foundation.html/.

5. John Steinbeck, *The Grapes of Wrath* (New York: Viking Penguin, 1939).

6. Wayne B. Solley, Robert R. Pierce, and Howard A. Perlman, "Estimated Use of Water in the United States in 1990," *U.S. Geological Survey Circular* 1081 (1993): 1–76.

7. Associated Press, "Oilman Looks into Selling West Texas Water. T. Boone Pickens identifies possible routes in marketing the region's newest commodity," *Corpus Christi Caller-Times* (May 18, 2000); this newspaper article available online at the Caller.com Web site, http://www.caller-times.com/2000/may/18/today/business/263.html/. For a provocative paper on the market of buying and selling freshwater, read Kenneth D. Frederick's *Marketing Water: The Obstacles and the Impetus*. "As water grows more precious," it announces, "so do the incentives—and the innovations—to try to apply market principles to its use and management." Resource Article 132 (Washington, D.C.: Resources for the Future, summer 1998), online at http://www.rff.org/environment/water.htm/.

8. See Solley, Pierce, and Perlman, "Estimated Use" (note 6).

9. Check out the *Green Power Switch* Web sites at http://www.TVA.gov/greenpowerswitch/green_mainfaq.htm/ and http://www.kub.org/.

10. See the references in note 4.

11. W. H. McAnnally, P. H. Burgi, D. Calkins, R. H. French, J. P. Holland, B. Hsich, B. Miller, and J. Thomas, "Water Resources," pages 261–290, in R. G. Watts, ed., *Engineering Response to Global Climate Change: Planning a Research and Development Agenda* (New York: CRC Press, Lewis Publishers, 1997).

12. Kenneth D. Frederick and Peter H. Gleick, "Water Resources and Climate Change," in N. J. Rosenberg et al., eds., *Greenhouse Warming: Abatement and Adaptation* (Washington, D.C.: Resources for the Future, 1989); Kenneth Frederick, *Water Resources and Climate Change*, Climate Issues Brief no. 3 (Washington, D.C.: Resources for the Future, June 1997), available online at Web site http://www.rff.org/environment/water.htm/.

13. Curt Suplee, "El Niño, La Niña, Nature's Vicious Cycle," *National Geographic* (March 1999), pp. 72–95. For the forecast predicted for the next El Niño event, check out the Web site http://www.nationalgeographic.com/elnino/.

14. In this 1998 electronic publication entitled "Children of the Tropic: El Niño and La Niña," Bob Henson and Kevin E. Trenberth provide an excellent synthesis of the history of these phenomena, of ocean linkages that interact with climate mechanisms, and of future forecasts for El Niño events. See the Web site at http://www.ucar.edu/communications/lasers/elnino/.

15. See Suplee, "El Niño" (note 13).

16. Barbara A. Miller and W. Gary Brock, *Sensitivity of the Tennessee Valley Authority Reservoir System to Global Climate Change*, Report WR28-1-680-101 (Norris, Tenn.: TVA Engineering Laboratory, 1988).

17. See note 16.

18. Barbara A. Miller, V. A. Alavian, M. D. Bender, D. J. Benton, P. Ostrowski, Jr., J. A. Parsly, H. M. Samples, and M. C. Shiao, *Impact of Incremental Changes in Meteorology on Thermal Compliance and Power System Operations*, Report WR28-1-680-109 (Norris, Tenn.: TVA Engineering Laboratory, 1992); Barbara A. Miller, V. A. Alavian, M. D. Bender, D. J. Benton, M. C. Shiao, P. Ostrowski, and J. A. Parsley, *Sensitivity of the*

TVA Reservoir and Power System to Extreme Meteorology, Report WR28-1-680-111 (Norris, Tenn.: TVA Engineering Laboratory, 1992).

19. Miller and Brock, *Sensitivity* (see note 16).

20. See note 9.

21. The online Lehman's Catalog for nonelectric products and eco-friendly home furnishings can be found at the Web site http://www.lehmans.com/.

22. Joel B. Smith, Rich Richels, and Barbara Miller, "Potential Consequences of Climate Variability and Change for the Western United States," in chapter 8 NAST, *Climatic Change Impacts* (see note 4).

23. Smith, Richels, and Miller, "Potential Consequences" (see note 22).

24. Nasser Bateni, Paul Hutton, Waiman Yip, and Bob Zettlemoyer, *Bulletin 160-98: California Water Plan Update* (Sacramento: Division of Planning and Local Assistance, California Department of Water Resources, November 1998). This is a report to evaluate Greenhouse-world options for meeting California's future water needs for agricultural, urban, and environmental water uses projected in the year 2020. See it online at http://rubicon.water.ca.gov/b160index.html/.

25. Smith, Richels, and Miller, "Potential Consequences" (see note 22).

26. NAST, *Climatic Change Impacts* (see note 4).

27. U.S. Environmental Protection Agency (EPA), *The Colorado River Basin and Climate Change* (Washington, D.C.: EPA, December 1993); see this report online at the Web site http://www.epa.gov/globalwarming/publications/impacts/co_basin.html/.

28. Rob Turner, "Honey, We're Moving to Henderson. The fastest growing city in America—Henderson, Nevada—is also quickly becoming the nation's top retirement destination. Want to visit?" *Money* vol. 29, no. 7 (July 2000): 108–113; Web site http://www.money.com/bestretirement/.

29. See note 28. Complementary municipal and commercial Web sites for Henderson, Nevada, can be viewed at http://www.tegnet.net/HendersonNevadaGateway/index.htm/ and http://www.henderson-nevada.com/news.html/.

30. See note 28.

31. Information on the Christmas bird count can be found at the National Audubon Society Web site at http://www.audubon.org/bird/cbc/index.html/.

32. Terry L. Root, "Environmental Factors Associated with Avian Distributional Boundaries," *Journal of Biogeography* 15 (1988): 489–505.

33. Terry L. Root, *Atlas of Wintering North American Birds* (Chicago: University of Chicago Press, 1988).

34. Audubon Web site (see note 31).

35. Terry L. Root and Stephen H. Schneider, "Can Large-scale Climatic Models be Linked with Multiscale Ecological Studies?" *Conservation Biology* vol. 7, no. 2 (1993): 256–270.

36. See the EPA *Global Warming* Web site at http://www.epa.gov/ globalwarming/impacts/birds/index.html/.

Chapter 8. Ecological Survival Kit

1. C. S. Holling, ed., *Adaptive Environmental Assessment and Management* (Chichester, UK: John Wiley & Sons, 1978). For an example of cross-impacts analysis applied to global climate change, from which table 8-1 was adapted, see the publication by M. Parry and T. Carter, "The Fifth Step: Assessing the Impacts," chapter 8, pages 95–136, in M. Parry and T. Carter, *Climate Impact and Adaptation Assessment: a Guide to the IPCC Approach* (London: Earthscan Publications Ltd., 1998).

2. To EPA's one-stop source for the *Climate Change Outreach Kit*, click on the Web site at http://www.epa.gov/globalwarming/publications/outreach/ index.html/.

3. Michael Williams, ed., *Climate Change Information Kit* (Washington, D.C.: Global Environmental Facility coordinated for the United Nations Development Programme and Environment Programme [UNDP-GEF], 1999), available from the Web site at http://www.undp.org/gef/ new/ccinfo.htm/.

4. Ron P. Neilson, I. Colin Prentice, B. Smith, T. Kittel, and D. Viner, "Simulated Changes in Vegetation Distribution under Global Warming," pages 429–456, in R. T. Watson, M. C. Zinyowera, and R. H. Moss, eds., *The Regional Impacts of Climate Change: An Assessment of Vulnerability,* a Special Report of IPCC Working Group II (Cambridge: Cambridge University Press, 1998).

5. Robert S. Thompson, S. W. Hostetler, Patrick J. Bartlein, and Katherine H. Anderson, *A Strategy for Assessing Potential Future Changes in*

Climate, Hydrology, and Vegetation in the Western United States, U.S. Geological Survey Circular 1153 (Washington, D.C.: USGS, 1998).

6. From the *Extreme Weather Sourcebook,* you can examine both geographic and temporal patterns of economic losses caused by tornadoes in the United States.

7. R. L. Ritschard, coordinator, *Draft White Paper Report for the Southeastern Regional Workshop* (Huntsville, Ala.: Global Hydrology and Climate Center, 1997); this 72-page report is available online at http://www.ghcc.msfc.nasa.gov/regional/assessment_national.html/.

8. See note 7.

9. John M. Connell, "Diversity in Tropical Rain Forests and Coral Reefs," *Science* 199 (1978):1302–1310.

10. Reid A. Bryson and F. Kenneth Hare, "The Climates of North America," pages 1–47 in Reid A. Bryson and F. Kenneth Hare, eds., *Climates of North America* (New York: Elsevier Publishing Company, 1974); Steward T. A. Pickett and Peter S. White, eds., *The Ecology of Natural Disturbance and Patch Dynamics* (New York: Academic Press, 1985); L. Brouillet and R. D. Whetstone, "Climate and Physiography," chapter 1, pages 15–46, in Nancy R. Morin, ed., *Flora of North America North of Mexico*, vol. 1, *Introduction* (New York: Oxford University Press, 1993).

11. Ritschard, *Draft White Paper* (see note 7).

12. Ken Dychtwald, *Age Power: How the 21st Century Will Be Ruled by the New Old* (New York: Jeremy P. Tarcher/Putnam, Penguin Putnam, 1999).

13. John C. Bogle, *Common Sense on Mutual Funds: New Imperatives for the Intelligent Investor* (New York: John Wiley, 1999).

14. Paul Ray identifies a major American subculture that he calls Cultural Creatives (see chapter 2). These CCs are considered thoughtful consumers that deliberately make socially responsible decisions concerning purchases and investments, based on their cultural, social, and environmental values. Three financial consultants, specializing in socially responsible investing, have published a key new book: Hal Brill, Jack A. Brill, and Cliff Feigenbaum, *Investing with Your Values: Making Money and Making a Difference* (New York: Bloomberg Press, 1999). Hal and Jack Brill have established Natural Investment Services, Inc.; their Web site at http://www.naturalinvesting.com/ evaluates and lists socially screened

mutual funds and funds investing in community development programs. Their listing procedure selects mutual funds based on issues of avoidance, affirmation, shareholder activism, and community improvement. Also, Cliff Feigenbaum publishes a quarterly newsletter, *The Green Money On-Line Guide and Journal* (http://www.greenmoney.com/), based in Spokane, Washington. This is an in-depth source of information for socially responsible investing, focusing on both socially and environmentally responsible businesses, investing opportunities, and green consumer resources. *Green Money*'s three goals of sustainability integrate equity with ecology and economy. The new magazine *Natural Business LOHAS Journal Tracking the Lifestyles of Health and Sustainability Markets*, and its Web site at http://www.LohasJournal.com/, concentrate on the $230-billion purchasing power of ecology-conscious Cultural Creatives.

15. Aldo Leopold, *A Sand County Almanac* (New York: Oxford University Press, 1966).

Chapter 9. Ten Best Strategies for Living Well in the Age of Global Warming

1. Clifton Leaf, "Fast Forward: What Retirement Will Look Like in the New Millennium," *Smart Money* (August 1999), pp. 114–121.

2–5. See Leaf, "Fast Forward" (note 1).

6. Sylvester J. Schieber and John B. Shoven, "The Economics of U.S. Retirement Policy: Current Status and Future Directions," pages 1–39 in Sylvester J. Schieber and John B. Shoven, eds., *Public Policy Toward Pensions: A Twentieth Century Fund Book* (Cambridge, Mass.: The MIT Press, 1997).

7. Richard C. Leone, preface, pages vii–ix in Schieber and Shoven, eds., *Public Policy Toward Pensions* (see note 6).

8. Ken Dychtwald, *Age Power: How the 21st Century Will Be Ruled by the New Old* (New York: Jeremy P. Tarcher/Putnam, Penguin Putnam, 1999).

9. Leaf, "Fast Forward" (see note 1).

10. J. Gertner and R. Kirwan, "The Best Places to Live '99." *Money* vol. 28, no. 11 (November 1999): 130–134. The Web site for *Money* (http://www.money.com/money/depts/retirement/bpretire/) helps you sort through the preliminaries for a new hometown. Pick your own balance of priorities (weather, environment, education, cost of living), tied to demographic trends

(health care, leisure, arts, crime, transportation, city size) for nearly five hundred metropolitan areas across the United States.

11. William Sterling and Stephen Waite, *Boomernomics: The Future of Your Money in the Upcoming Generational Warfare* (New York: The Library of Contemporary Thought, The Ballantine Publishing Group, 1998).

12. Garrison B. Keillor, *Lake Wobegon Days* (New York: Viking Penguin, 1990).

13. Katherine Pearson, "From the Editor, The migration of *Coastal Living* from issue to issue." *Coastal Living* (September-October 1999) vol. 3 (Issue 5): 20. Check out this Web site for coastal lifestyles at http://www.coastalliving.com/.

14. Representative Web site clearinghouses for time-shares can be found at http://www.timeshares.com/ and, for the largest company for exchanging vacation time-shares, Resort Condominiums International (RCI) at http://www.rci.com/ctg/cgi-bin/RCI/home/.

15. The Web site for the Good Sam Club, the oldest, largest organization for RV travel, is at http://www.goodsamclub.com/.

16. See the virtual tour of Paul Revere's famous ride from Boston in 1775 on the Web site for Boston National Historical Park, at http://www.nps.gov/bost/ftrail.htm/.

17. Mariners' dependence on electronic LORAN-C transmitters for navigation is quickly being broadened with international development of the Differential Global Positioning Service. The DGPS navigation system uses a combination of ground-based stations as well as the GPS array of satellites to accurately fix locations for recreational boats. For more information on these types of navigational beacons, see Web sites at the National Capital Freenet (http://www.ncf.carleton.ca/boating/dgps.html/), the U.S. Coast Guard Navigation Center (http://www.navcen.uscg.mil/), and the Canadian Coast Guard (http://www.ccg-gcc.gc.ca/tosd-dsto/awti/DGPS/main.htm/).

18. Harry S. Dent, Jr., *The Roaring 2000s: Building the Wealth and Lifestyle You Desire in the Greatest Boom in History* (New York: Simon and Schuster, 1998).

19. See EPA's global warming report entitled *Climate Change and Hawaii* at Web site http://www.epa.gov/globalwarming/impacts/stateimp/hawaii/index.html/.

Chapter 10. Legacy for Future Generations

1. See the discussion of creating a buffer zone for the Great Smoky Mountains National Park at the Foothills Conservancy Web site http://www.foothillsland.org/index.htm/.

2. Jonathon P. Ambrose and Susan P. Bratton, "Trends in Landscape Heterogeneity along the Borders of Great Smoky Mountains National Park," *Conservation Biology* vol. 4, no. 2 (1990):135–143.

3. For a discussion of the changes in species diversity and other aspects of landscapes as they are transformed from natural to cultural, see the book by Richard T. T. Forman and Michel Godron, *Landscape Ecology* (New York: John Wiley & Sons, 1985).

4. See note 3.

5. For a discussion of the strategy for positioning nature reserves to allow for adjustments in species ranges with global climate change, see R. L. Peters and J. D. S. Darling, "The Greenhouse Effect and Nature Reserves," *BioScience* 35 (1985):707–717.

6. See note 5.

7. See note 5.

8. An excellent discussion of the relationship of buffer zones and migration corridors to the design of nature reserves is contained in the Web site for the Wildlands Project: http://www.wildlands.org/corridor/reserve.html/.

9. See note 8.

10. Al Gore, *Earth in the Balance* (New York: Houghton Mifflin, 2000).

11. David M. Gates, *Climate Change and Its Biological Consequences* (Sunderland, Mass.: Sinauer Associates, 1993).

12. See note 3.

13. This statement was taken from a 1955 book describing the origins and role of the Nature Conservancy, and quoted within the book by Victoria M. Edwards, *Dealing in Diversity: America's Market for Nature Conservation* (Cambridge: Cambridge University Press, 1995). To find more information on how you can become involved in The Nature Conservancy, click on http://www.tnc.org/. For information about the activities of professional ecologists in the Ecological Society of America, see http://www.sdsc.edu/ESA.

14. The origins and present-day structuring of land trusts in the United States are described in detail in the book by Victoria Edwards, *Dealing in*

Diversity (see note 13). A number of land trusts and regional conservancies are profiled on the World Wide Web. For example, an electronic link to the Little Traverse Conservancy, as well as addresses for many additional conservancies and land trusts, is located at http://landtrust.org/ and http://www.landtrust.org/ltnames.htm/.

15. See note 1.

16. See note 1.

17. Fact sheets on biological diversity, as well as white papers on global climate change and land use management written by committees within the Ecological Society of America, can be found online at http://esa.sdsc.edu/climate.htm/.

18. The full text of Jane Lubchenco's presidential address to the *American Association for the Advancement of Science (AAAS)* can be found in: Jane Lubchenco, "Entering the Century of the Environment: a New Social Contract for Science," *Science* 279 (1998): 491–497. To read this paper online, search the category of science online at the Web site of the American Association for the Advancement of Science: http://www.aaas.org/.

19. For Joan Iverson Nassauer's design plan for ecological corridor neighborhoods, see the following publications: Joan Iverson Nassauer, A. Bower, K. McCardle, and A. Caddock, *The Cambridge Ecological Corridor Neighborhood: Using Ecological Patterns to Guide Urban Growth* (Minneapolis: University of Minnesota, 1997); Joan Iverson Nassauer, "Cultural Sustainability," pages 65–83 in Joan Iverson Nassauer, ed., *Placing Nature: Culture and Landscape Ecology* (Washington, D. C.: Island Press, 1997); Joan Iverson Nassauer, "Culture as a Means for Experimentation and Action," pages 129–133 in John A. Wiens and Michael R. Moss, eds., *Issues in Landscape Ecology* (Snowmass Village, Colo.: International Association for Landscape Ecology, Fifth World Congress, 1999).

20. The material for this section came primarily from a newspaper article entitled "How Does the U.S. Fish and Wildlife Service Feel about Salt Water Hammocks? (We know how developers and commissioners feel about them)," *Golden Isles Weekend* no. 95 (April 2–8, 1999). See it on the Web at http://www.weekendonline.net/. For general insights into issues in the conservation of biological diversity, see Edward O. Wilson and D. L. Perlman, *Conserving Earth's Biodiversity: An Interactive Learning Experience for Studying Conservation Biology and Environmental Science* (Cov-

elo, Calif.: Island Press, 1999). Also check out http://www.islandpress.org/wilsoncd/.

21. See the Web site of the Environmental Protection Agency (EPA) on *Green Communities Program,* outlining what citizens can do to work with local planners to develop greenways for native plant and animal species, for better designs for conserving water resources, and for restoration of damaged wetlands and other environmentally sensitive areas: http://www.epa.gov/region03/greenkit/index.html/.

22. Rick Dubrow, Stacy Fehlenberg, and John Peine, "SAMAB Initiative Highlights: Sustainable Communities," *Southern Appalachian Man and the Biosphere,* SAMAB News (2000), p. 2. Check the progress of this pilot program on the *Sustainable Communities Initiative* at http://samab.org/.

23. The Dewees Island Web site contains a list of "sustainable development quick facts" that have been used as a basis for landscape planning for this upscale resort community: visit on the Web site at http://www.deweesisland.com/culture_facts.asp/.

Resources

Reference Books

Botkin, Daniel B., *Discordant Harmonies, A New Ecology for the Twenty-first Century* (Oxford: Oxford University Press, 1990).

Dent, Harry S. Jr., *The Roaring 2000s Investor: Strategies for the Life You Want* (New York: Simon and Schuster, 1999).

Edwards, Victoria M., *Dealing in Diversity: America's Market for Nature Conservation* (Cambridge: Cambridge University Press, 1995).

Gates, David M., *Climate Change and Its Biological Consequences* (Sunderland, Mass.: Sinauer Associates, 1993).

Houghton, John T., *Global Warming: The Complete Briefing, Second Edition* (New York: Cambridge University Press, 1997).

Peters, R. L., and T. E. Lovejoy, eds., *Global Warming and Biological Diversity* (New Haven, Conn.: Yale University Press, 1992).

Schneider, Stephen H., *Global Warming: Are We Entering the Greenhouse Century?* (San Francisco: Sierra Club Books, 1989).

Sterling, William, and Stephen Waite, *Boomernomics: The Future of Your Money in the Upcoming Generational Warfare* (New York: Library of Contemporary Thought, Ballantine Books, 1998).

Thompson, Robert S., S. W. Hostetler, Patrick J. Bartlein, and Katherine H. Anderson, *A Strategy for Assessing Potential Future Changes in Climate, Hydrology, and Vegetation in the Western United States*. U.S. Geological Survey Circular 1153 (Washington, D.C.: USGS, 1998).

Web Site Resources

1. For an up-to-date listing of Greenhouse-world related web sites, check our Web site at **http://www.boomerbreakpoints.com/**.

2. U.S. National Assessment Team (NAST), *Climatic Change Impacts for the United States: The Potential Consequences of Climatic Variability and Change, Foundation Document* (Washington, D.C.: U.S. Global Change Research Program [USGCRP], 2000), available at the Web site http://www.gcrio.org/NationalAssessment/foundation.html/. The summary reports are presented in an Overview Document and in a more in-depth Foundation Document (Web site http://www.gcrio.org/NationalAssessment/foundation.html/).

This newest national assessment has been developed within the U.S. Global Change Research Program (USGCRP; see Web sites http://www. usgcrp.gov/ and http://www.gcdis.usgcrp.gov/pubs.nap.html/). Substantive reports are being prepared for twenty key geographic regions (for the status of these reports, check http://www.nacc.usgcrp.gov/regions/): Alaska, Appalachians, California, eastern Midwest, Great Lakes, Great Plains (central, northern, southern), Gulf Coast, Hawaii and Pacific Islands, metropolitan East Coast, Middle Atlantic, Native Peoples/Native Homelands, New England, Pacific Northwest, Rocky Mountains and Great Basin, South Atlantic Coast and Caribbean, Southeast, Southwest (Colorado River Basin, Rio Grande River Basin). Each of these regional syntheses provides Greenhouse-world projections for the next twenty-five to thirty years, covering the Boomer Breakpoint in 2025, and for the year 2100.

3. The *Global Warming* Site from the Environmental Protection Agency is found at http://www.epa.gov/globalwarming/impacts/stateimp/index.html/. For EPA's one-stop source for the *Climate Change Outreach Kit*, click on the Web site at http://www.epa.gov/globalwarming/publications/outreach/index. html/. The EPA *Green Communities Program* is dedicated to developing sustainable green communities, that is, communities that "integrate a healthy environment, a vibrant economy, and a high quality of life" at http://www. epa.gov/region03/greenkit/index.html/.

4. The Federal Emergency Management Agency (FEMA) projects areas likely to be inundated by Greenhouse-world floods as part of its mandate through the National Flood Insurance Program (http://www.fema.gov/nfip/).

5. Louis R. Iverson, Anantha M. Prasad, B. J. Hale, and E. K. Sutherland, *An Atlas of Current and Potential Future Distributions of Common Trees of the Eastern United States*, General Technical Report NE-265, (Delaware, Ohio: Northeastern Research Station, USDA Forest Service, 1999). An interactive electronic version of this atlas is available online at the Web site http://www.fs.fed.us/ne/delaware/atlas/index.html/.

6. A number of land trusts and regional conservancies are profiled on the World Wide Web. For example, the electronic link to the *Little Traverse Conservancy*, as well as addresses for many additional conservancies and land trusts, is located at http://landtrust.org/ and http://landtrust.org/ltnames.htm/.

7. The National Aeronautics and Space Administration (NASA) *Global Change Master Directory* (GCMD) can be reached online at the Web site http://gcmd.gsfc.nasa.gov/.

8. The PEW Center of Global Climate Change presents a wide-ranging suite of authoritative position papers. These excellent syntheses characterize the state-of-the-science concerning climate change, as well as providing provocative reports on the kinds of Greenhouse-world dangers that threaten coastlines, agricultural production of food, and freshwater supplies (Web site http://www.pewclimate.org/projects/).

9. James Titus and Vijay Narayanan, *The Probability of Sea Level Rise* (Washington, D.C.: Environmental Protection Agency [EPA], 1995). For an online copy of this report (#EPA 230-R95-008), see the Web site at http://www.epa.gov/globalwarming/publications/impacts/sealevel/probability/chapt_9.pdf/.

10. Coordinated as a public service outreach by the U.S. Global Change Research Program (USGCRP), the *Gateway to Global Change Data and Information System* (GCDIS) can be found at the Web site http://globalchange.gov/.

Index

A

Acadia National Park, ME, 200
Acid Deposition Standard Feasibility Study, 93, 200
acid rain, 88–95, 106, 136, 199–200
Adirondack Mountains, NY, 90, 93
affluence line, 15
Age of Global Warming, xi–xv
Age Wave, 10, 13, 18–19, 39, 163
air pollution, 88–96
Alberta Clipper, 76–79, 165
alien invaders, xiii, 7, 107, 127, 174
Asheville, NC, 88
"Ask Dr. Global Change," 142
Antarctic Ice Cap, 45
asset allocation, 145–46
Atlantic Coast (Seaboard), 42, 44–45, 47, 52–55, 141, 160, 162, 166, 190, 192
Atlantic Ocean, 49–53, 190

B

Babbit, Bruce, 176
Baby Boomer Generation, xii–xiii, 3–4, 10–23, 77, 106, 124–25, 129, 150, 168, 181–82
back of beyond, 8, 88
balsam fir, 84

barrier islands, 60–64, 178
Base Flood Elevation, 57–61
bats, 108–109, 203
beech, 80–83, 103
bicoastal strategy, 160, 210
Big Bend National Park, TX, 200
biological diversity (biodiversity), 80–82, 86, 96–97, 136, 138, 159, 167, 171–73, 177, 181, 197, 201, 212
biosphere reserves, 95, 171–173, 181, 211
bird watching, 122 125
black bear, 169, 171
Blue Ridge Mountains, 88, 93, 98
blueberries, 84–85, 198
Blue Ridge Parkway, 88, 91
Bogle, John, 145–46, 208
Boomer Breakpoints, xiii, 12–14, 20–23, 30, 32–34, 39, 41, 43, 45, 50–51, 53–56, 61–63, 66–67, 77, 84–85, 93, 98, 105–106, 115–16, 119, 122, 129–130, 141, 146–47, 156, 160, 162, 182, 201
Boomer demographics, 11–14, 19–23, 41, 120–22, 145, 147, 183
Boomer Economics (Boomernomics), 11, 18–19, 21–22, 154, 183–85, 210, 214

Boomer Initiative Web Site, 15, 184
Botkin, Daniel B., 82, 84–86, 197–198, 214
Boundary Waters Canoe Area (BWCA), MN, 84
Bourgeois Bohemian (Bobo) subculture, 15–16, 18, 184
Bradley, Ray, 28, 186
Bratton, Susan, 100, 170, 202, 211
Brock, W. Gary, 113–17
brook trout (brookies), 89, 93–95, 199–200
Brooks, David, 15–16, 184
brown trout, 93–94
buffer zones, 170, 173, 211
business-as-usual (BAU) scenario (of economic growth), 31–34, 39, 43

C

California Water Plan, 120, 206
Cambridge, MN, 177–179
Canadian Climate Centre, 68, 194
carbon dioxide gas (CO_2), 25–26, 30, 32, 34, 43
Caribbean islands, 41, 167
Cash Flow King, 154–55, 157–59
catastrophe, 49–55, 60–64, 115, 139–141, 156, 174

Census Bureau, U.S., 10, 150, 183
Center for Limnology, 70
Chabot, Brian, 99, 202–203
Chattanooga, TN, 112, 114
Cherokee, NC, 95, 170
chlorofluorocarbon (CFC) gas, 27, 32, 43, 186
Christmas bird count, 122, 206
"civilized rustic lifestyle," 151
climate change (global projections), 6, 12, 27–36, 66–68, 129, 138, 193
Climate Change Information Kit, 138
Climate Change Outreach Kit, 137, 207, 215
Climate Data Library, 137–138
Coastal Barrier Resource Act, 56
coastal erosion, 62–64, 180, 191
Coastal Resource Atlases, 62, 192
cold-hardiness (of plants), 80, 98, 196–97
Colorado Low, 76–79
Colorado River Basin, 120, 141, 206
comfort seeker, xiv, 143, 154–55, 159–64
community planning, xiii, 19, 64, 117–19, 136, 148, 169–71, 174, 177–179, 181
conservation easement, 176
conservation ethic, 147
contingencies, 132–42
core area, 172–73
corridors, 86–87, 103, 170, 173, 181, 211–12
Corps of Engineers, U.S. Army, 58

Coutant, Charles, 54–55, 191
cross-impacts analysis, 132, 173, 207
Crowley, Tom, 28–29, 186, 194
Cultural Creative subculture, 16–17, 184, 208
custom hazard map, 141

D

Davis, Margaret B., 81–82, 197
"deal with it" (adapt), 58, 81
deciduous forest, 80, 99, 101, 195, 202–203
decisions, 142–43
deep time, 42, 79–80
deforestation, 32, 34, 94
Delcourt, Hazel R., 195–97, 201, 203
Delcourt, Paul A., 196–97, 201, 203
Dent, Harry S., Jr., 10–11, 167, 183–84, 210, 214
Department of Natural Resources (DNR), 56, 87, 198
Dewees Island, SC, 180, 213
dieback (of forests), 82, 86–87, 98, 100–101, 105, 144, 167
dot-com generation, 11, 17, 129
Dow-Jones stock index, 146, 152
Druyan, Leonard, 51
Duffy, D. C., 100, 202
dust-bowl droughts, 35, 68, 108–109, 166
Dychtwald, Ken, 10, 152–53, 183, 208–209

E

eco-entrepreneur, 86, 155–56
eco-futurist, xiv, 80, 139, 159, 182

Ecological Contrarian, 19, 154–56
ecological edge, 4, 7, 181–82
ecological legacy, 4, 9, 60–64, 130, 169–82
ecological literacy, 19, 180
ecological loser, 155, 164
Ecological Nomad, 119, 122–25, 155, 161, 163–64
ecological refugee, 107–109, 130, 147, 155, 158–59
ecological risk, 60–64, 153–55, 168
ecological security, xiii, 148–49, 167
ecological set-aside, 170
Ecological Society of America, 174, 177, 211–12
Ecological Speculator, 154–57
ecological stakeholder, 36, 97–98, 106, 131, 141–42, 158
ecological survival kit, 22–23, 129, 131–50, 182, 207
ecological survivor, 155, 164
Ecological Thrill Seeker, xiv, 143, 154–59
ecological vulnerability, xiv, 112–22, 144, 148–49, 155
economic growth (global projections), 31–34
ecotone, 103–106
Edwards, Victoria, 174, 211, 214
Eichenlaub, Van, 76, 78, 195
electrical brownouts and blackouts, 107, 116–19, 165–66
electrical grid, life off of, 8, 118
El Niño, 112–19, 136, 142, 168, 205

Emanuel, K. A., 51–52, 60–61, 190, 192
Encarta Terraserver Site, 139
endangered vistas, 91–93
endemic species, 94–95, 104
engineered retreat (of coastal construction), 59–60
enlightened self-interest, 19, 182
Environmental Protection Agency (EPA), 44, 58, 93, 98, 117, 120, 124, 137–38, 140, 168, 180, 189, 191–92, 194, 200, 206–207, 210, 213, 215–16
Environmental Systems Research Institute, 141
extinction, 7, 56–60, 81–86, 100, 102–106, 167
Extreme Weather Sourcebook, 140, 208

F

Federal Emergency Management Agency (FEMA), 48, 56–58, 61–62, 119, 131, 140–41, 157–58, 190–91, 215
financial risk, 145–46
financial security, 145–46
first alert, 124
fishery stocks, 66, 71–74, 94–95, 142
Flood Insurance Rate Maps (FIRMs), 56–57, 140
flooding, 56–57, 80, 113–17
Florida peninsula, 52–55, 64, 173
fly fishing, 89, 93–95, 200
Foothills Conservancy, TN, 170, 175–76, 211

forest fragmentation, 34, 94–95, 100–101, 170–74, 178
forest stand simulation model (JABOWA), 82–86
forestry practices, 86, 198
fossil fuel energy, 27, 33–34, 90, 111, 117, 136
Fourier, Jean-Baptiste, 25
"four bite rule," 55–58
Four Corners area (of U.S. Southwest), 77
Frank, Neil, 46
Franklin, Jerry, 198
Fraser fir, 98, 106
Frederick, Kenneth, 110, 204
fresh water (as a critical resource), 110, 204
fresh water fish, 66, 71–74, 93–95, 194
"fun-or-flight syndrome," 150-51

G

Galveston, TX, 58–60
Gates, Bill, 14–15, 184
Gates, David M., 27, 34–36, 174, 185–86, 188, 211, 214
Gateway to Global Change Data and Information System (GCDIS), 138, 216
Gatlinburg, TN, 95–96, 170–71
Geological Survey, U.S., 139–40, 196, 204
general circulation models (GCMs) (for simulating future climate), 27–30, 66–68, 192–93, 207
Generation-X (Gen-Xers), 11, 14–15, 17, 129, 151–52, 154–56
Gertner, Jon, 15–16, 184, 209–210
GFDL climate model (full-boil version of

future global warming), 30, 32, 54–55, 67, 72, 74, 81–83, 102–105, 114, 187, 193
GISS climate model (slow-simmer projection of future global warming), 30, 32, 51–52, 54–55, 67, 72, 74, 81–86, 102–105, 113–17, 187, 190
glaze-ice storms, 68, 78, 144, 165
Global Change and Climate History Project, 138, 207–208, 214
Global Change Master Directory (GCMD), 137, 216
Global Change Research Program (USGSCRP), U.S., 138, 141, 188–89, 215–16
Global Climate Change Site, 138
Global Environmental Facility of the United Nations (UNDP-GEF), 138, 207
Global Hydrology and Climate Center (GHCC), 142, 208
global warming, 6, 12, 24–36, 75
Global Warming Site, 98, 124, 138, 207, 215
Golden Isles, GA, 178
golfing, 8, 107, 134, 181
Good Sam Club, 163, 210
Gore, Al, 93, 211
"gray gold," 15, 156
Great Lakes Environmental Research Laboratory, 71
Great Lakes region, 65–78, 81–86, 139, 141, 155, 159–60, 165–67, 193–95, 197

Great Smoky Mountains
National Park, TN-
NC, 89–106, 160, 162,
169–72, 175, 199–202,
211
Greeley, Horace, 107
green communities,
177–180, 212–13
*Green Communities
Program*, 180, 213,
215
green power, 36, 111, 166
*Green Power Switch
Program* (for renewable
energy sources), 111,
118, 205
Green scenario (of
economic growth),
33–34
Greenhouse effect, 25–26
Greenhouse gases, 12,
25–27, 30, 34–36
Greenhouse loser, xiv,
106, 130, 155
Greenhouse winner, xiv,
106, 130, 155, 168,
182
Greenhouse-world
climate, 6, 12, 24–36
Greenhouse-world
lifestyle chart, 133, 135
Greenland Ice Cap, 45
groundwater, 108–110
Gulf Coast, 42, 45, 53,
64, 81, 141, 162
Gulf of Mexico, 47, 49,
52, 77, 111, 114, 160,
167
Gulf Stream, 52–54

H
habitat contrast, 170–73,
211
Hansen, Jim, 30, 187
Hawaii, 141, 168, 210
hemlock, 81–83, 89, 101,
103
Henderson, NV, 121–22,
206
Hengeveld, Henry, 68–69,
194

Henson, Bob, 113, 205
Hicks, David, 99,
202–203
Holling, C. S., 132, 207
Homesteader, 155,
165–67
Houghton, Sir John,
24–27, 30, 185–87,
214
hunting, 8, 66, 86–87
hurricane categories,
46–49, 60–62
hurricane protective
shields, 52–53
hurricane temperature
threshold, 46, 48–55
hydroelectric power,
111–19

I
ice age, 28, 42, 104–105,
108, 195, 197
impact cost (for
expanding municipal
infrastructure), 148
index mutual funds,
145–46, 208
individual retirement
accounts (IRAs), 12
Industrial Revolution, 7,
26–27
Intergovernmental Panel
on Climate Change
(IPCC), 24–25, 32–34,
43–44, 139, 185, 187,
192
intermediate disturbance
hypothesis, 144, 208
Iverson, Louis, 103, 203,
216

J
jack pine, 84
Jones, Landon, 10, 183

K
Kephart, Horace, 88, 199
Key West, FL, 164
Kirtland's warbler, 84,
178
Knoxville, TN, 111, 169

L
La Niña, 65, 102, 113–19,
136, 142, 168, 205
Lake Erie, 72, 75
Lake Huron, 65, 69, 75,
79, 165, 167
Lake Michigan, xii,
65–66, 69, 71–75, 77,
165, 167, 195
Lake Nipigon, 155
Lake Ontario, 75, 165,
167
Lake Superior, 65, 69, 72,
75, 77, 79, 165, 167
Lake Wobegon, 155, 210
lake trout, 66, 71–74
lake-effect climate, 75,
77–78, 82, 87, 165,
167, 195
lakefront real estate,
69–70, 82, 136, 148
Lakeshore lifestyle destination, 65–87
land trust, 175–76, 181,
211–12, 216
landscape planning, 134,
169–71, 180–81,
211–12
Lansing, MI, 75, 78
large-mouth bass, 71, 74
Leaf, Clifton, 150–51,
209
leaf peepers (of fall
foliage), 8, 89,
101–106
Leone, Richard, 152
Leopold, Aldo, 147, 209
lifestyle destinations,
39–40
lifestyle factors, 133–37
Little Traverse Conser-
vancy, MI, 176, 212,
216
"living off the electrical
grid," 8, 118
"lose it" (go extinct), 7,
58–60, 81–86, 100,
102–106, 167
Lower Peninsula of
Michigan, 65–66, 68,
79, 176, 192

Lubbock, TX, 109
Lubchenco, Jane, 177, 212

M
McCormick, Michael, 71–72, 195
McCracken, Gary, 108–109, 203
McFadden, Daniel, 20–22, 146, 185
Mackinac Bridge, MI, 65–66, 75, 192
Magnuson, John, 70–74, 195
Manabe, Syokuro, 30, 187
marine fish, 54
maritime forest, 178, 212
Matterhorn melt-down model (of plunging home values), 20–22, 146–47
Meadows, Donella, 31, 186
Meier, Albert, 100, 202
Mesa Water, Inc., 110, 204
methane gas, 26–27
Miami, FL, 42, 60
Michener, William, 52, 190–91
midwestern United States, 77–78, 92, 141, 147, 165, 193–94
Millennium Kids, 11, 14, 17, 129
Miller, Barbara, 113–17, 189, 205–206
Minneapolis-St. Paul, MN, 177
Mitchell, Joni, 170
Modern subculture, 16–18
Moore, Steve, 89, 94
Mount LeConte, TN/NC, 91, 96, 98, 104, 201
"move it" (migrate to a more suitable environment), 58–60, 81–82, 98, 103–106, 124–25, 197

Mountain lifestyle destination, 88–106
mushroom hunting, 87, 198
Muskegon, MI, 75, 77–78

N
Nash, Steve, 98, 199, 202
Nassauer, Joan Iverson, 177–79, 212
National Aeronautic and Space Administration (NASA), 30, 187, 190, 216
National Assessment, U.S., 141, 188-89, 193-94, 197, 206, 215
National Audubon Society, 122, 206-207
National Center for Atmospheric Research (NCAR), 53, 193
National Flood Insurance Program (NFIP), 56–57, 140, 215
National Geophysical Data Center (NGDC), 139
National Hurricane Center (NHC), 46–48, 190
National Oceanographic and Atmospheric Administration (NOAA), 30, 71, 139, 187–88, 190
National Park Service (NPS), 92, 94, 200
National Weather Service (NWS), 46
Nature Conservancy, 174–75, 181, 211
nest egg, 145–47, 159, 161
New England Climate Initiative (NECI), 200
New England region, 52, 54–55, 67–68, 103–104, 141, 155, 165–66

New Orleans, LA, 58
New Social Contract for Science, 177, 212
nitrogen oxide gas, 26–27, 90, 92
nor'easter storms, 165
northern Appalachian Mountains, 98, 104–105, 166–67, 200
northern white cedar, 85
Norton, D. C., 75, 195
nuclear power, 111, 117

O
oak, 80, 98, 101, 103
Oak Ridge National Laboratory, TN, 54
ocean front real estate, 47–49, 60–64, 136, 156–57
Ogallala Aquifer, 108–109, 203–204
old-growth forest, 82, 86, 95, 98, 100
"once-in-a-hundred-years" flood, 56
overturn (of lake-water circulation), 72
ozone gas, 26–27, 90
ozone hole, 27
ozone layer, 27
ozone pollution, 27, 65, 89–91, 138, 186, 199–200

P
Pacific Coast, 44, 119
Pacific Northwest, 141, 160, 165, 198
Pacific Ocean, 49, 113, 168
paleoecologist, xii, 79–80, 104–105, 159, 195–96, 201, 203
Pearson, Katherine, 160, 210
Perpetual Beach Walker, 155, 160–61
Peters, R. L., 82, 85, 196–97, 214

PEW Center for Global Change, 30, 44, 53, 67, 139, 188, 191, 193, 216
pH, 89, 94
Pickens, T. Boone, 110, 204
Pigeon Forge, TN, 95, 170
pocket wilderness, 96, 105
population growth (global projections), 31–34
priorities for Greenhouse-world lifestyles, 8, 9, 131–32, 150–53, 159
property insurance, 53–58, 119
property tax assessment, 147–48

Q
quaking aspen, 85
quality of life, 4, 5, 18, 31, 88–89, 96, 155, 168

R
rainbow trout, 93–94
Raper, S. C. B., 33, 43, 49–50, 187–89
Ray, Paul, 16–17, 184, 208
Recreational Vehicles (RVs), 8, 147, 149, 162–63, 210
red maple, 84, 101, 103
red pine, 84
Renfro, Jim, 90–91, 199
retirement assets, 21–22, 151–53, 183–84
retirement communities, 121–22, 181, 206, 209–210
retirement projections (for Boomers), 12–14, 17–23
risk management, 142
risk-takers, 153–54
river flow, 113–20

Rocky Mountains, 77, 108, 119, 141, 173
Rodionov, S. N., 76, 195
rolling easements (for engineered retreat of threatened coastal homes), 60
Root, Terry, 122–24, 207
Rustbelt, 21, 122, 147

S
S & P 500 stock index, 146
safe sites, 18, 107, 163, 165
Saffir, Herbert, 46
Saffir-Simpson scale (of hurricane categories), 46–49, 190
Saint Lawrence Seaway, 69, 165, 194
salmon, 66, 71, 73–74
San Francisco, CA, 44
sassafras, 90, 103
save-our-savings (SOS) strategy, 157
Schieber, Sylvester, 21–22, 185, 209
Schlesinger, M. E., 67, 190, 192–93
Schneider, Stephen, 27, 124, 185–87, 207, 214
sea-breeze convection zones, 166
Seafaring Vagabond, 155, 163–64
sea islands of South Carolina and Georgia, 62, 178, 192
sea-level rise, 35, 42–46, 58–64, 188–92, 216
Seaside lifestyle destinations, 41–64
Shoven, John, 21–22, 185, 209
Simpson, Robert, 46
SLOSS (Single Large or Several Small reserves) dilemma, 171
Snowbird, 8, 122, 124–25, 155, 159–60

social risk, 146–48
social security, 146–48
Social Security Administration, 13–14, 67, 152, 183–85
socially responsible investing, 146, 208–209
Solley, Wayne, 204
song birds, 122–23, 171, 174, 178
South Florida Building Code (for Miami-Dade County), 48, 190
southeastern United States, 52–53, 84, 92, 108, 141–42, 144, 159, 190, 208
Southern Appalachian Assessment (SAA), 89, 199
Southern Appalachian Man and the Biosphere (SAMAB), 97, 180, 199, 201–202, 213
Southern Appalachian Mountain Initiative (SAMI), 91, 199
southern Appalachian Mountains, 89–92, 96, 100, 103–104, 114, 162, 180, 199
Southern High Plains, 67, 141, 147, 204
Southern Nevada Water Authority (SNWA), 121–22
southern pine, 103
southwestern United States, 108, 119–22, 141, 160
Special Flood Hazard Areas, 56, 140
species extinctions, 7, 58–60, 81–86, 94–95, 100, 102–106, 167
species migrations, 58–60, 81–82, 98, 103–106, 124–25, 171–73, 197
sports fishing, 8

Spring Wildflower
Pilgrimage, 97, 162,
202
spruce-fir forest, 96, 98,
104–106
stability seeker, xiv, 143,
154–55, 164–68
Steinbeck, John, 109
Sterling, William, 11,
21–22, 183–85, 210,
214
stewardship (of the land),
64, 174–75, 181–82
storm surge (associated
with hurricanes), 42,
46–49, 60–64, 164,
180
strategies for Greenhouse-
world lifestyles, 150–68
structure (of lake ecosystems), 70–73
sugar maple, 81–83, 85,
101, 103
sulfur dioxide gas, 89,
92–93
Sunbelt lifestyle destination, 8, 107–125, 148
Sun-Seeking Road
Warrior, 155, 162–63
sustainability, xiii, 16, 31,
96, 118, 136, 176–81,
198, 212–13
Sustainable Hedonist,
155, 167–68

T
"target repetitive loss
buildings" (FEMA), 57,
191
tax boards, 147–48
Teal, Edwin Way, 97,
202–203
Tennessee River Basin,
111–12
Tennessee Valley
Authority (TVA),
111–19, 205–206
Thatcher, Margaret,
24–25, 185
thermal habitat (within
lakes), 70, 195

thermocline, 72
Thompson, Robert S.,
196, 207–208, 214
timber harvest, 80, 84,
85, 94, 100–101, 198,
202
time-share condo-
miniums, 8, 98, 161,
210
Time-Share Timer,
97–103, 155, 161–62
Titus, James, 44–45, 58,
188–89, 191–92, 216
Toffler, Alvin, 17, 184
topographic maps,
139–40
TopoZone Site, 140
Tornado Alley, 140, 165
tourists' paradox, 95–96
Traditional subculture,
16–17
treeline, 80, 104
trillium, 94–100, 169
tulip tree, 90
tundra, 80, 104–105
turkey, 66, 87

U
Upper Peninsula of
Michigan, 65–66, 75,
79, 148, 192

V
Vegetation/Ecosystem
Modeling and Analysis
Project (VEMAP), 138,
207
voluntary simplicity, xiii,
9, 147, 153

W
Waite, Stephen, 11, 21–22,
183–85, 210, 214
walleye, 71, 73–74
Washington, D.C., 45
water wars, 110–12
wealth factor, 14
Weatherald, Richard, 30,
187
White, Peter, 104,
201–203, 208

white birch, 84
whitefish, 71, 73–74
white pine, 84, 90
white-tailed deer, 66, 87
Wigley, Tom, 30, 33,
42–43, 53, 67, 186–88,
191, 193
wildfire, 35, 80, 84, 87,
144, 198
wildflower life strategies,
99–100
wildflower pilgrim, 8, 89,
96–101
wildflower species,
96–101, 169, 201–202
Wildlands Project, 173,
211
Wilshire 5000 stock
index, 146
wolf, 171
World Bank, 32–34
World out of Control
scenario (of economic
growth), 34
www.boomerbreak-
points.com, xv, 132,
195, 215

Y
Year-Rounder, 155,
164–68
yellow birch, 81–83, 103
yellow perch, 71, 73–74
Yosemite National Park,
CA, 200
Yooper, 66, 75

CHELSEA GREEN

Sustainable living has many facets. Chelsea Green's celebration of the sustainable arts has led us to publish trend-setting books about organic gardening, solar electricity and renewable energy, innovative building techniques, regenerative forestry, local and bioregional democracy, and whole foods. The company's published works, while intensely practical, are also entertaining and inspirational, demonstrating that an ecological approach to life is consistent with producing beautiful, eloquent, and useful books, videos, and audio cassettes.

For more information about Chelsea Green, or to request a free catalog, call toll-free (800) 639-4099, or write to us at P.O. Box 428, White River Junction, Vermont 05001. Visit our Web site at www.chelseagreen.com.

Chelsea Green's titles include:

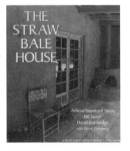  

The Straw Bale House
The New Independent Home
The Natural House
Serious Straw Bale
The Beauty of
    Straw Bale Homes
Treehouses
The Resourceful Renovator
Independent Builder
The Rammed Earth House
The Passive Solar House
Wind Energy Basics
Wind Power for Home &
    Business
The Solar Living Sourcebook
A Shelter Sketchbook
Mortgage-Free!
Stone Circles
Toil: Building Yourself

The Neighborhood Forager
Gaia's Garden: A Guide to
    Home-Scale Permaculture
The Apple Grower
The Flower Farmer
Breed Your Own
    Vegetable Varieties
Passport to Gardening
The Soul of Soil
The New Organic Grower
Four-Season Harvest
Solar Gardening
Straight-Ahead Organic
The Contrary Farmer
The Co-op Cookbook
The Bread Builder
Simple Food for the
    Good Life
The Maple Sugar Book

Believing Cassandra
This Organic Life:
    Confessions of a Suburban
    Homesteader
Gaviotas: A Village to
    Reinvent the World
Who Owns the Sun?
Global Spin
Hemp Horizons
A Place in the Sun
Beyond the Limits
The Man Who Planted Trees
The Northern Forest
The New Settler Interviews
Loving and Leaving the
    Good Life
Scott Nearing: The Making
    of a Homesteader
Wise Words for the Good Life